Springer Theses

Recognizing Outstanding Ph.D. Research

For further volumes:
http://www.springer.com/series/8790

Aims and Scope

The series "Springer Theses" brings together a selection of the very best Ph.D. theses from around the world and across the physical sciences. Nominated and endorsed by two recognized specialists, each published volume has been selected for its scientific excellence and the high impact of its contents for the pertinent field of research. For greater accessibility to non-specialists, the published versions include an extended introduction, as well as a foreword by the student's supervisor explaining the special relevance of the work for the field. As a whole, the series will provide a valuable resource both for newcomers to the research fields described, and for other scientists seeking detailed background information on special questions. Finally, it provides an accredited documentation of the valuable contributions made by today's younger generation of scientists.

Theses are accepted into the series by invited nomination only and must fulfill all of the following criteria

- They must be written in good English.
- The topic should fall within the confines of Chemistry, Physics, Earth Sciences, Engineering and related interdisciplinary fields such as Materials, Nanoscience, Chemical Engineering, Complex Systems and Biophysics.
- The work reported in the thesis must represent a significant scientific advance.
- If the thesis includes previously published material, permission to reproduce this must be gained from the respective copyright holder.
- They must have been examined and passed during the 12 months prior to nomination.
- Each thesis should include a foreword by the supervisor outlining the significance of its content.
- The theses should have a clearly defined structure including an introduction accessible to scientists not expert in that particular field.

Dai-Ming Tang

In Situ Transmission Electron Microscopy Studies of Carbon Nanotube Nucleation Mechanism and Carbon Nanotube-Clamped Metal Atomic Chains

Doctoral Thesis accepted by
Institute of Metal Research,
Chinese Academy of Sciences, China

Author
Dr. Dai-Ming Tang
Shenyang National Laboratory
for Materials Science
Institute of Metal Research
Chinese Academy of Sciences
Shenyang
People's Republic of China

Supervisors
Prof. Hui-Ming Cheng
Prof. Chang Liu
Shenyang National Laboratory
for Materials Science
Institute of Metal Research
Chinese Academy of Sciences
Shenyang
People's Republic of China

ISSN 2190-5053 ISSN 2190-5061 (electronic)
ISBN 978-3-662-52411-4 ISBN 978-3-642-37259-9 (eBook)
DOI 10.1007/978-3-642-37259-9
Springer Heidelberg New York Dordrecht London

Softcover reprint of the hardcover 1st edition 2013

Printed on acid-free paper

Springer is part of Springer Science+Business Media (www.springer.com)

Parts of this thesis have been published in the following journal articles:

Tang DM, Liu C, Cheng HM (2006) Platelet boron nitride nanowires. NANO 1: 65–71.

Tang DM, Liu C, Cheng HM (2007) Controlled synthesis of quasi-one-dimensional boron nitride nanostructures. J Mater Res 22: 2809–2816.

Tang DM, Liu C, Li F, Ren WC, Du JH, Ma XL, Cheng HM (2009) Structural evolution of carbon microcoils induced by a direct current. Carbon 47: 670–674.

Tang D-M, Yin L-C, Li F, Liu C, Yu W-J, Hou P-X, Wu B, Lee Y-H, Ma X-L, Cheng H-M (2010) Carbon nanotube-clamped metal atomic chain. Proc Natl Acad Sci USA 107: 9055–9059.

Liu B, Tang D-M, Sun C, Liu C, Ren W, Li F, Yu W-J, Yin L-C, Zhang L, Jiang C, Cheng H-M (2011) Importance of Oxygen in the Metal-Free Catalytic Growth of Single-Walled Carbon Nanotubes from SiO_x by a Vapor–Solid–Solid Mechanism. J Am Chem Soc 133: 197–199.

Tang DM, Zhang LL, Liu C, Yin LC, Hou PX, Jiang H, Zhu Z, Li F, Liu BL, Kauppinen EI, Cheng HM (2012) Heteroepitaxial Growth of Single-Walled Carbon Nanotubes from Boron Nitride. Scientific Reports 2: 971.

Supervisors' Foreword

Carbon nanotube (CNT) is a representative one-dimensional nanomaterial with unique structure and attractive physical chemical properties. Enormous efforts and vast investments have been devoted to the research and development of CNTs in the past two decades. Notable progresses have been made on the preparation, property, and application exploration of CNTs. However, there are still challenges remained for the research community, i.e., the structure-controlled synthesis, growth mechanism, and industrial application of CNTs.

Aiming at these above critical subjects, Dr. Dai-Ming Tang's doctoral thesis focuses on revealing the growth process and mechanism of CNTs by using in situ TEM technique. A novel approach including the preparation of two-end open CNT nano furnace, selective loading of catalyst nanoparticles inside the CNT hollow core, Joule heating of the CNT furnace by applying a voltage under TEM, carbon supply by electron beam irradiation, and the observation and study of the growth of CNTs from nanoparticles was first developed. By using this technique, it was found that the growth of CNTs from non-metal SiO_x catalyst follows a vapor–solid surface–solid mechanism, quite different from the traditional vapor–liquid–solid mechanism for metal catalyst, and that it is possible to grow CNTs from a pure carbon system, without adding any metal or non-metal catalyst. These results will be helpful to get a proper understanding of the growth mechanism of CNTs, and hence facilitate their structure-controlled growth.

Dr. Tang also designed and prepared a CNT-clamped metal atomic chain structure. By using a metal-filled CNT as starting material, the outer carbon layer and inner metal nanorod were finely machined under TEM, and finally a CNT-clamped metal atomic chain was fabricated. The forming process, structure, and quantized electrical transport property of this hybrid structure were studied and correlated. This work provides an alternative way for the connection and assembly of metal atomic chains, and demonstrates that CNTs can be used as electrodes connecting nano/subnano structures in nanodevices due to their excellent electrical and mechanical properties as well as desired stability.

It has been demonstrated in this doctoral thesis that it is an efficient way to study the growth mechanism and device fabrication of CNTs by using in situ TEM. It is expected that more valuable results can be achieved by combining this technique with theoretical calculations and controlled growth of CNTs closely in the next stage.

Shenyang, January 2013

Hui-Ming Cheng
Chang Liu

Acknowledgments

My research in this thesis was done under the supervision of my respectable mentors, Prof. Hui-Ming Cheng and Prof. Chang Liu. I would like to take this opportunity to show my gratitude to them. Professor Cheng has profound knowledge in carbon sciences. He is a serious scholar with great passion and motivation for innovation. He keeps on thinking and rethinking of the development trends in the research field. At every stage of my research, Prof. Cheng provided valuable instructions and inspiring suggestions. Professor Liu led me into the gate of science and encouraged me when I achieved every small improvement. He gave me the freedom to have my own thoughts and we often discussed about the latest advances in research. He contributed a lot to this thesis, from the selection of the topic, to the experimental design, data analysis, manuscript writing and revisions, and so on.

I would like to give my special gratitude to Prof. Feng Li. He has wide interests, quick and penetrating thoughts with unique visions. During the past 6 years, Prof. Li and I had a lot of fruitful discussions, from which I learned to think differently. As we both like photography, on the weekends, we often went out into the wild to take photos of the nature and had random discussions about undefined topics, truly enjoyable journeys.

My research could not be possible without the strong support from Prof. Xiu-Liang Ma. He offered professional instructions for the microstructure analysis, and allowed me sufficient time to complete my challenging experiments.

I want to express my appreciation to my great collaborators, Dr. Li-Chang Yin, Dr. Cheng-Hua Sun, Mr. Bilu Liu, Mr. Wan-Jing Yu, Dr. Peng-Xiang Hou, Dr. Qing-Feng Liu, Ms. Chun-Xiang Shi, Dr. Rui-Tao Lv, Dr. Jin-Hong Du, Mr. Jing-Bao Cheng, and Ms. Li-Li Zhang. Through the collaborations with Dr. Yin, I expanded my knowledge of physics greatly. Dr. Sun has the ability to find out the essence of things in an effective and direct way. He provided effective support to me for the understanding of the experimental results. I learned a lot from Mr. Liu about clear and innovative thinking. Dr. Hou and Mr. Yu provided high quality samples for my in situ TEM experiments by smart design of AAO template methods. Dr. Liu, Ms. Shi, and Dr. Lv generously offered metal-filled CNTs. I thank Dr. Du and Mr. Cheng for the interesting carbon microcoils. And I also thank Ms. Zhang for her great contributions to the synthesis and characterizations of SWCNTs by using BN nanofibers as growth seeds. In addition, I would like to present special acknowledgments to Dr. Oleg Lourie from Nanofactory. We did many interesting experiments together. He often helped me out of experimental troubles and provided useful instructions to make sure the experiments going smoothly.

During the 6 years in SYNL, I received professional and kind support from the technicians; they are Mr. Kui-Yi Hu, Mr. Bo Wu, and Dr. Chuan-Bin Jiang, Dr. Jun Tan, Mr. Xiao-Ping Song, Ms. Shu-Hua Yang, Ms. Yu-Zhen Sun, Ms. Bin Zhang, Ms. Wei Gao, and Mr. Fei-Xue Yang. They helped me in various experiments, and structural characterizations.

I thank the family-like support from the advanced carbon materials division. And importantly, I would like to thank the teachers and administrators of the graduate school for providing me with a peaceful and active environment for living and studying, and thank my classmates for the experiences of learning and research they shared with me, which will be my unforgettable memories.

For the publication of current translated thesis, I would like to thank my Editor, Ms. June Tang for her professional work, kind and patient help.

Last but not least, I want to express my deepest gratitude to my family for their generous support. Especially, I appreciate the great sacrifice of my wife; she was and will continue to be my best friend in my lifetime.

Contents

Abbreviations

AFM	Atomic force microscopy
BCC	Body-centered cubic
CCD	Charge-coupled device
CNT	Carbon nanotube
CVD	Chemical vapor deposition
DFT	Density functional theory
DOS	Density of states
EDS	Energy-dispersive X-ray spectroscopy
EELS	Electron energy loss spectroscopy
FCC	Face-centered cubic
FET	Field-effect transistor
G_0	Conductance quantum
HCP	Hexagonal close-packed
HRTEM	High-resolution transmission electron microscopy
MAC	Metal Atomic Chain
MBE	Molecular beam epitaxy
MCBJ	Mechanically controllable break junction
MD	Molecular dynamics
MEMS	Microelectromechanical systems
SAED	Selected area electron diffraction pattern
SPM	Scanning probe microscopy
STM	Scanning tunneling microscope
TEM	Transmission electron microscopy
VLS	Vapor–liquid–solid
VSSS	Vapor–solid surface–solid

Chapter 1
Introduction

Carbon-based advanced materials are important for the modern civilizations, such as carbon composite materials for airplanes, graphite electrodes for steels manufacturing, and graphite as moderator for nuclear power reactors [1, 2]. According to the electronic structures and hybridization, traditional carbon materials could be classified as either sp^2 hybridized graphite or sp^3 hybridized diamond. In recent years, zero-dimensional fullerenes [3], one-dimensional carbon nanotubes (CNTs) [4], and two-dimensional graphenes [5] have become new members of the carbon family. They are ideal model materials for low-dimensional sciences, and are regarded as the key materials for future nanoscience and nanotechnology [2, 6–9].

Since the discovery in 1991 [4], CNTs have attracted great interest in the fields of physics, chemistry, materials science, and biology because of the unique structure, excellent mechanical, physical and chemical properties, and potential applications in electronics, mechanics, and energy [10–12]. In the following sections, recent progresses in CNT research will be introduced, including the controllable synthesis, growth mechanism, electronic devices, and in situ transmission electron microscopy (TEM) investigations. After that, the motivation, and organization of this thesis will be presented.

1.1 Progresses in Controllable Synthesis of CNTs

A major research direction is the controllable synthesis of CNTs, because of the close relationship of structure and properties [13], as shown in Fig. 1.1. Early explorations were mainly focused on the developments of new synthesis methods, increase of production scale, and improvements of purity. There are three main methods for the production of single-walled CNTs (SWCNTs): arc-discharge [14, 15], laser ablation [16], and chemical vapor deposition (CVD) [17, 18]. Currently, CVD is the most popular method because of the low cost, convenient parameter control, and possibility to scale up. Depending on the catalyst loading methods, CVD method could be classified into different types, such as floating catalyst [17], loaded catalyst [19] and surface growth [18], and so on.

D.-M. Tang, *In Situ Transmission Electron Microscopy Studies of Carbon Nanotube Nucleation Mechanism and Carbon Nanotube-Clamped Metal Atomic Chains*, Recognizing Outstanding Ph.D. Research, DOI: 10.1007/978-3-642-37259-9_1,

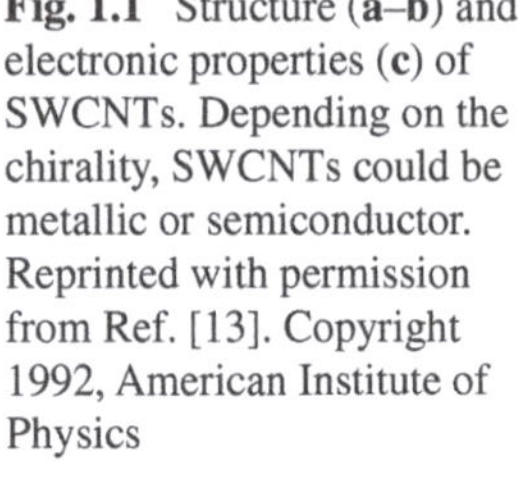

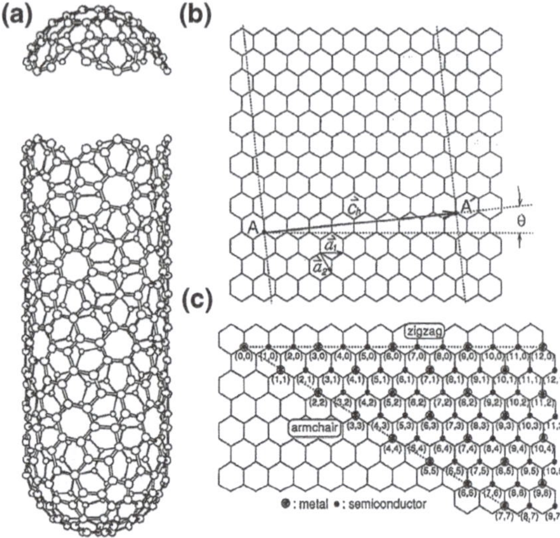

Fig. 1.1 Structure (**a**–**b**) and electronic properties (**c**) of SWCNTs. Depending on the chirality, SWCNTs could be metallic or semiconductor. Reprinted with permission from Ref. [13]. Copyright 1992, American Institute of Physics

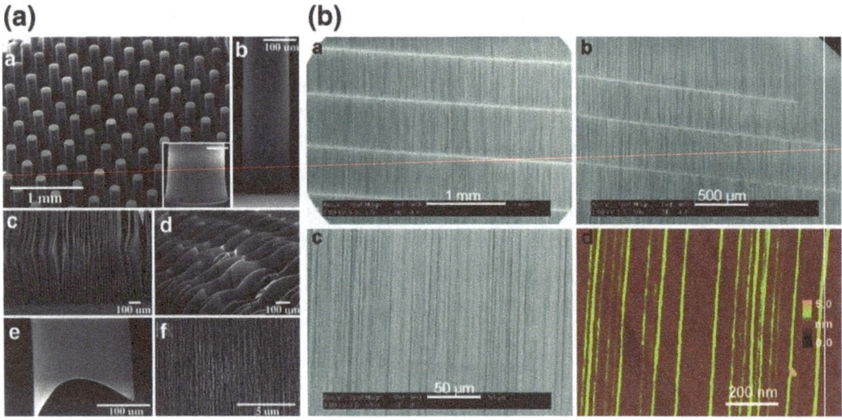

Fig. 1.2 **a** SEM images of SWCNT forests by super growth, from Ref. [24]. Reprint with permission from AAAS. **b** High density SWCNT horizontal array. Reprinted with the permission from Ref. [28]. Copyright 2008, American Chemical Society

In recent years, the controllable synthesis of CNTs by CVD method has made great progresses. For example, the growth position could be controlled by the position of the catalysts, and the growth direction could be controlled by applying electrical field, gas flow, or lattice matching techniques [20–22]. As a result of the alignment control, CNT arrays in both horizontal and vertical directions have been produced. In addition, the length of the CNTs could be tuned by controlling the catalyst lifetime and activities [21, 23]. Figure 1.2 presents two representative works. In 2004, Hata group from National Institute of Advanced Industrial Science and Technology (AIST) reported the super growth of high purity SWCNT arrays by using a water-assisted CVD technique [24]. Currently, mass production of SWCNTs forests with this technique has been realized, and the applications of SWCNTs in microelectromechanical systems (MEMS), energy storage, etc., are under development [25–27]. Another example is the high density horizontal arrays developed by Liu group from Duke University in 2008. A density as high as 20/μm of aligned SWCNTs was achieved by a lattice control technique [28].

The applications in electronic devices require CNTs with uniform electronic properties which are controlled by the chirality. Therefore, a lot of efforts were made to control the chirality and conductance [29–36]. For example, Resasco et al. reported the synthesis of SWCNTs with enriched chiralities by using a Co–Mo catalyst (Fig. 1.3) [29]. Li et al. and Chiang et al. realized certain selectivity of

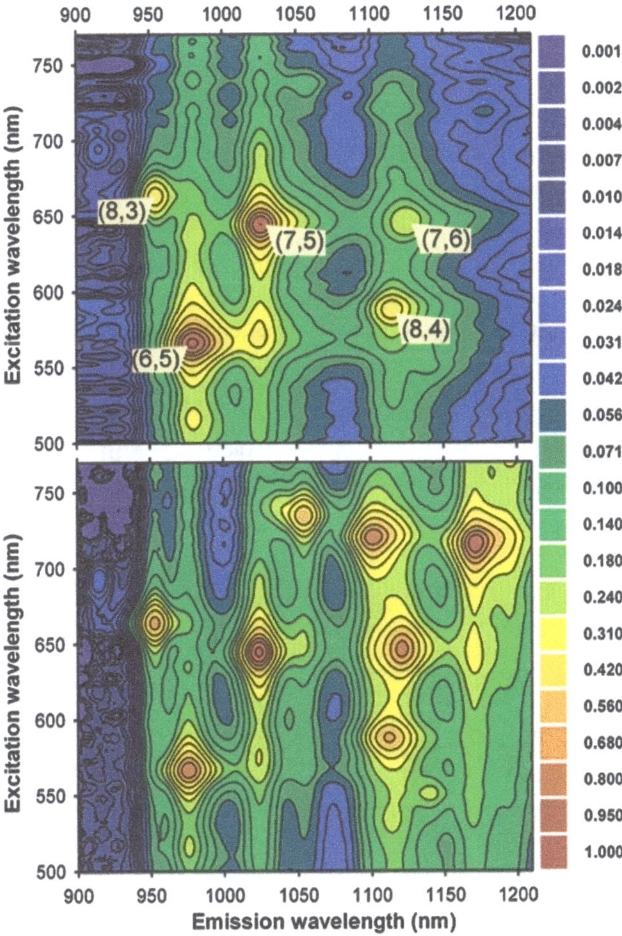

Fig. 1.3 Photoluminescence emission intensities for the Co–Mo SWCNTs with a narrow (n, m) distribution. Reprinted with the permission from Ref. [29]. Copyright 2003, American Chemical Society

chiralities by optimizing the catalyst composition, size, and growth temperatures in Fe–Ru and Fe–Ni systems, respectively [33, 37]. In addition, metallic enriched SWCNTs were produced by Harutyunyan et al. [35], by tuning the atmosphere during the pretreatments of the catalysts and thus influencing the shape of the catalyst particle and possibly their interaction with carbon.

To summarize, the key to the discovery of SWCNTs is the use of suitable catalyst. The control of macro arrangements of CNT assemblies is based on the manipulation of the size, density, and position of catalysts. In addition, the conductivity and chirality distribution could be tuned by optimizing the composition and pretreatment of the catalysts. Therefore, the key to realize the controllable synthesis of CNTs is the understanding of the growth mechanism, especially the role of the catalyst. In this thesis, the state of the catalysts during CNTs growth is investigated by using in situ TEM techniques, so as to reveal the CNTs growth mechanism (Chap. 3).

1.2 Progresses in Understanding CNTs Growth Mechanisms

One of the well-accepted growth mechanisms for one-dimensional materials is the vapor–liquid–solid (VLS) mechanism [38]. It was first proposed by Wagner and Ellis in 1964 to explain the growth of silicon whiskers (Fig. 1.4). The VLS

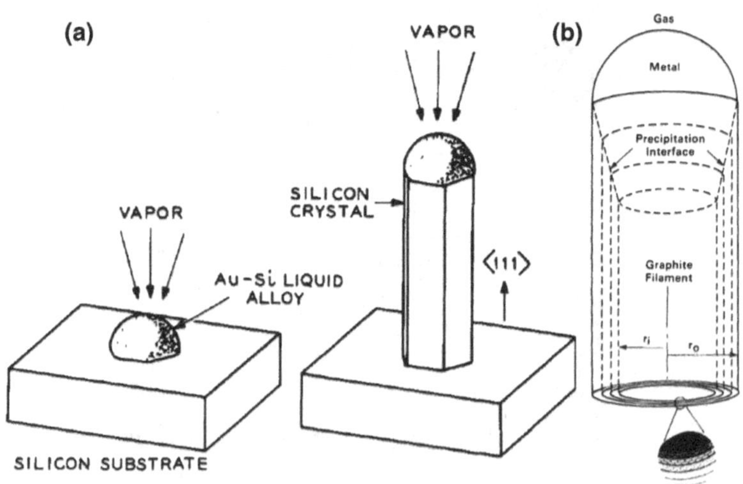

Fig. 1.4 **a** VLS growth mechanism of silicon whiskers. Reprinted with the permission from Ref. [38]. Copyright 1964, American Chemical Society. **b** Growth model of carbon filaments following the VLS mechanism. Reprinted with permission from Ref. [40]. Copyright 1984, American Institute of Physics

mechanism includes several steps. First, vaporized silicon precursors (V) are dissolved into catalyst particles (Au) to form a eutectic phase with a lower melting point and therefore the particle becomes liquid (L). The concentration of silicon increases until over saturation is reached. After that, solid silicon phase precipitates from the liquid catalyst particle (S). As the precipitation continues, a silicon nanowire is nucleated and grown. Baker and Tibbetts adopted the VLS mechanism to explain the growth of carbon fibers [39, 40]. The proposed growth process included the decomposition of carbon precursors, adsorption on the catalyst surface, dissolution into the catalyst particle, diffusion through the particle, and precipitation after oversaturation.

After the discovery of CNTs, various techniques were used to investigate the growth mechanism, such as the high-resolution transmission electron microscopy (HRTEM) and molecular dynamics (MD) simulations. Dai et al. proposed a *yarmulke* growth model based on the HRTEM observations of the relationship of SWCNTs and catalyst particles at different growth stages (Fig. 1.5) [41]. It was suggested that a carbon cap was precipitated from the catalyst particle and lifted up for the nucleation of a SWCNT. And then the SWCNT grows following the VLS mechanism.

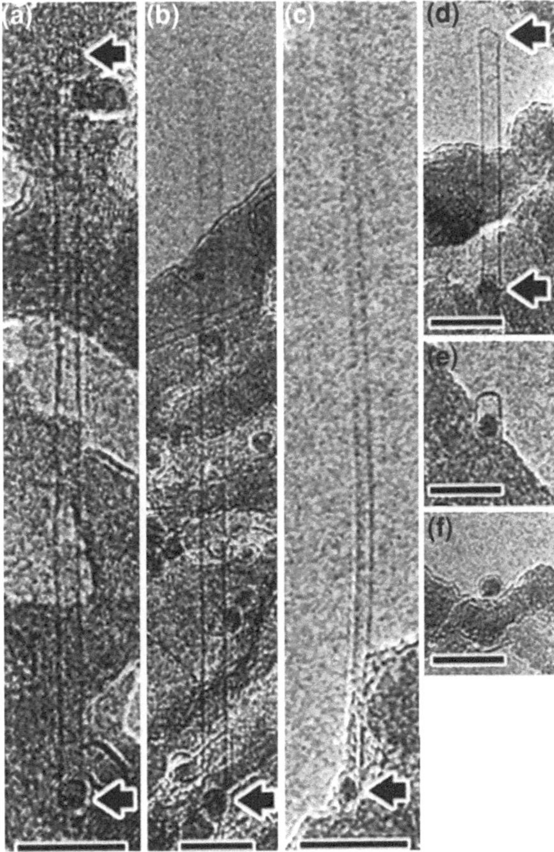

Fig. 1.5 Relationship of catalyst particles and SWCNTs by HRTEM, representing the *yarmulke* growth mechanism. Reprinted with the permission from Ref. [41]. Copyright 2001, American Chemical Society

A new method was developed in our group in 1998 for the mass production of SWCNTs by using a floating catalyst CVD technique [17]. It was found that sulfur was an efficient growth promoter, which enhances the growth efficiency and influences the diameter distribution. HTREM characterizations revealed, interestingly, no direct dependence between the size of the catalyst particles and the diameter of the grown nanotubes. The catalyst particle was usually larger than the nanotube. Therefore, a growth mechanism was proposed to explain the growth of SWCNTs by the floating catalyst CVD method (Fig. 1.6). Because of the adsorption of sulfur at preferred positions, the catalyst particle becomes locally liquid. The liquid zones have higher activities for the nucleation and growth of CNTs. Therefore, the diameter of the CNT is determined by the size of the liquid zone rather than by the whole particle [42–44].

The growth process of CNTs is complicated and influenced by many factors such as the catalyst size and composition, temperature, chemistry of the carbon source, and the crystalline structures of the substrates. What's more, the effects of these factors are correlated rather than independent, making the growth mechanism even more difficult to be understood. In addition, the structure of both the catalyst and CNTs could be changed during and after the growth. Therefore, it is not enough to study only by parameter optimization combined with post characterizations to reveal the CNTs growth mechanism. Therefore theoretical calculations and simulations were carried out to simulate the growth processes, including the growth positions (root or tip growth model), kinetics, nucleation process and diffusion route (bulk or surface diffusion), carbon–metal interactions, and so on [45–48]. For example, Ding et al. studied the interactions of many different kinds of metals

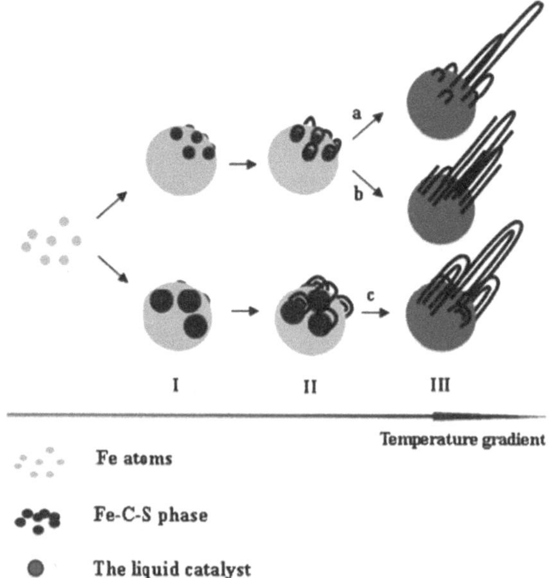

Fig. 1.6 Schematic of the growth mechanism of SWCNTs by floating catalyst CVD method. Reprinted with the permission from Ref. [44]. Copyright 2006, American Chemical Society

with carbon shells (Fig. 1.7). It was found that only when the interaction is strong enough, the carbon cap could be lifted up for the nucleation of a nanotube [48].

Traditionally, the catalysts for CNTs growth were mainly Fe-group metals and alloys. However, recently many new metal catalysts were discovered, including Au, Ru, Cu, Mn, Pt, Pd, Mo, Cr, Sn, Au, Mg, Al, Pb, and so on [49–54]. Most of these metals could not form carbide; therefore the growth mechanism of CNTs from the new metal catalysts should be different from traditional VLS mechanism. A modified VLS mechanism involving a size-dependent dissolution of carbon in nano-sized metals was proposed by Homma et al. (Fig. 1.8) [52].

Besides the new metal catalysts, surprisingly, metal-free catalysts were discovered for the CNTs growth, including semiconductors such as Si, Ge, and SiC, oxides such as SiO_x and ZrO_2, nano-carbon materials such as diamond, C_{60}, and open SWCNTs [55–61]. In addition, catalyst-free growth of CNTs on substrates was reported [62]. Because these metal-free catalysts usually have high melting points, a vapor–solid surface-solid (VSSS) growth mechanism was proposed by Homma et al. (Fig. 1.9). According to the proposed mechanism, carbon atoms are provided by surface diffusion on the catalyst surface rather than bulk diffusion through the particle. Up to now, there are no clear answers to many important and fundamental questions, such as the composition of the active catalysts, the diffusion route of carbon atoms, and so on.

To summarize, although the growth mechanism of CNTs growth has been investigated intensively, there are still many unsolved problems, especially after the discovery of new metal and metal-free catalysts. Therefore, in this thesis, a direct in situ TEM technique is used to investigate the growth mechanism, focusing on the state of catalyst during CNTs growth.

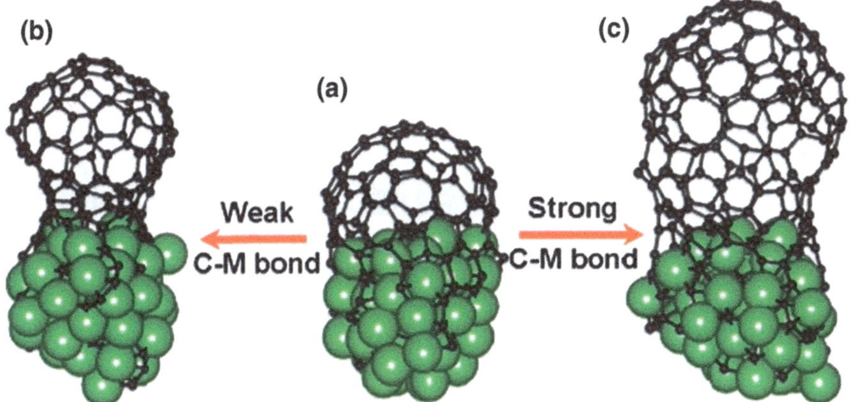

Fig. 1.7 Effects of carbon–metal interaction on the nucleation process of SWCNTs by molecular dynamics (MD) simulations. Reprinted with the permission from Ref. [48]. Copyright 2007, American Chemical Society

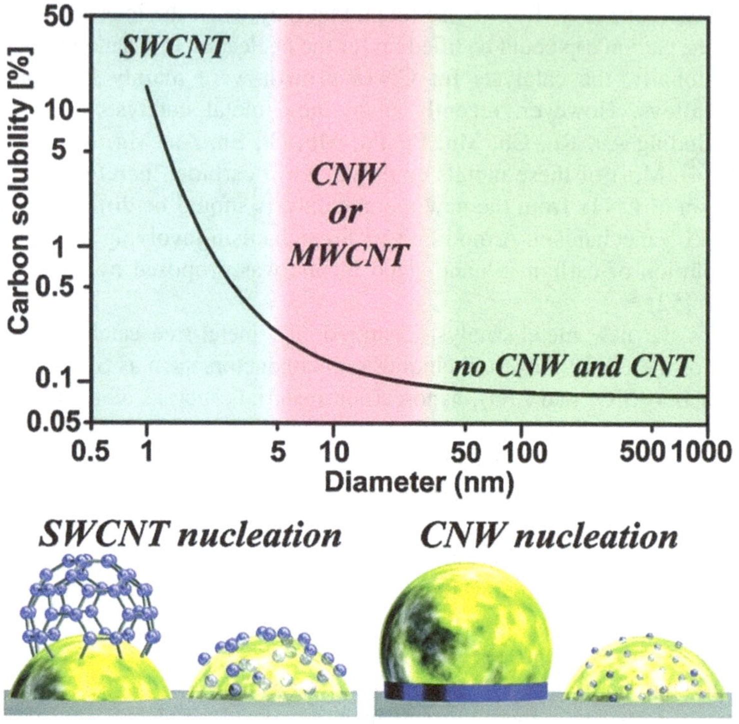

Fig. 1.8 Size-dependent carbon solubility in metal particles and nucleation models of carbon nanofibers and nanotubes from gold nanoparticles. Reprinted with the permission from Ref. [52]. Copyright 2008, American Chemical Society

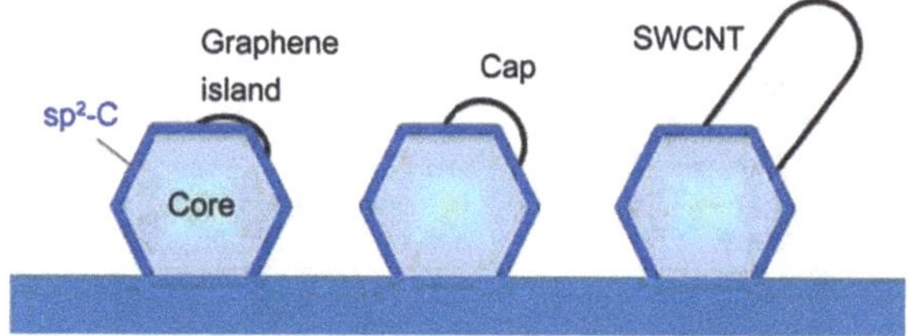

Fig. 1.9 VSSS growth mechanism proposed by Homma et al. Reprinted from Ref. [63], with kind permission from Springer Science Business Media

1.3 Progresses in CNTs-Based Nano-Electronic Devices

One of the main potential applications of CNTs is the nano-electronic devices because of the excellent electronic properties. Depending on the chirality, SWCNTs may behave as metals or semiconductors, with distinct applications.

1.3.1 CNTs as One-Dimensional Metallic Wires

Because of the nanosized diameter and large aspect ratio, the electron transport in CNTs becomes ballistic, which could be described by Landau equations, with free diffusion path approaching the micron scale [64, 65]. In addition, CNTs are bonded by carbon–carbon covalent bonds which is the strongest in nature, so CNTs are highly resistant to electromigration with the maximum current up to 10^{10} A/cm^2, 1000 times higher than that of copper wires. Therefore, CNTs could be used as high current cables with low power consumption [66]. Mann et al. measured the electronic transport of a single SWCNT by using Pd electrodes for Ohmic contacts (Fig. 1.10) [67]. Room temperature quantized conductance was reported based on the tests results, with the free diffusion path up to 500 nm at room temperature and 4 μm at low temperature [68]. Postma et al. manipulated a single SWCNT by atomic force microscopy (AFM) to form a kink. Coulomb blockade through the kink was measured, showing potential applications in single electron transistors [69]. Superconducting properties were reported by Tang et al. in the small diameter SWCNTs fabricated from zeolite templates, with the critical temperature up to 15 K [70]. Tsukagoshi et al. reported the coherent spin transport through the interface

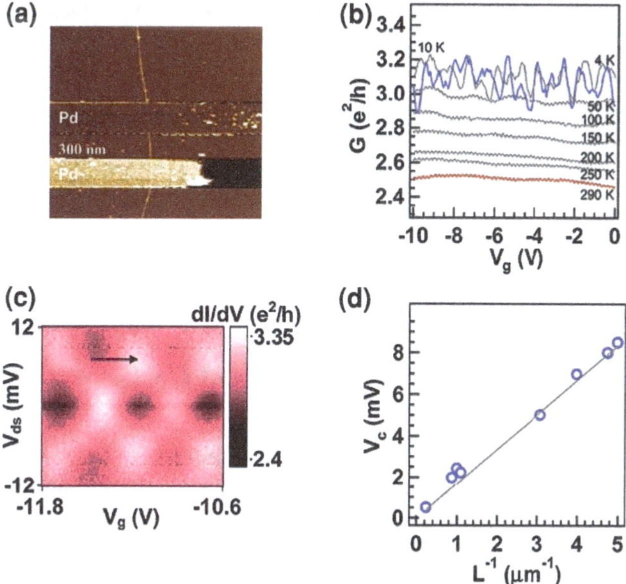

Fig. 1.10 Atomic force microscopy (AFM) image (**a**), and electrical conductivity measurements on a SWNT contacted by Pd (**b–d**), demonstrating ballistic transport. Reprinted with the permission from Ref. [67]. Copyright 2003 American Chemical Society

between ferromagnetic metals such as cobalt and CNTs, with the magnetoresistance effects up to 9 %, showing promising applications in nano-spintronics [71].

1.3.2 CNTs as Semiconductors

Compared with traditional semiconductors, CNTs have the advantage of symmetric band structure for electrons and holes, and similar mobilities for both charge carriers, therefore they are expected to have superior performance for nanoscale complementary metal–oxide–semiconductor (CMOS) devices [72]. In addition, CNTs have direct band gaps therefore making the combination of microelectronics and optoelectronics possible in the same material [73].

Martel et al. and Tans el al. first reported the CNT field effect transistors (FETs) in 1998 by using noble metals such as Au and Pt as electrodes [74, 75]. The CNTs had p-type characteristics and the on-off ratio was as high as 10^5. Soh et al. used transitional metals as the electrodes and effectively reduced the contact resistance from MΩ to kΩ range [76]. And Wind et al. improved the gate techniques and lowered the gate voltages from 12 V to about 0.5 V [77]. Martel et al. and Bockrath et al. reported the n-type CNT FET fabricated by vacuum annealing, alkali metal doping, and suitable selection of electrode materials [72, 78]. Based on these techniques, Derycke et al. built logic circuits [79]. In addition, because of the small diameter and large aspect ratio, CNTs have excellent optical transmittance and flexibility, which make them potential candidates for fabricating transparent and flexible devices (Fig. 1.11) [80].

As already mentioned, semiconducting CNTs have important potential applications. However, currently the synthesis of CNTs with uniform structure and conductivity still remains a big challenge. The applications of CNTs as conducting wires do not have so strict requirements and the synthesis of few-walled CNTs is becoming mature. Therefore, in this thesis, a new CNT-based nano-electronic device is designed by using CNTs as nanoscale clamps for the extremely small atomic chains (Chap. 4).

1.4 Progresses of In Situ TEM Studies of CNTs

It is learned from the history of science that the length scale of our study reflects the depth of our understanding of the world. For example, the invention of telescope extended our view deep into the space and back to the beginning of the universe. On the other direction, the microscopes enable us to look into the fascinating small world. TEM provided structural information of the materials with atomic resolution, and became one of the most powerful tools to study the structure-properties relationship of materials. In recent years, in situ TEM techniques were developed by introducing various attachments, such as gas, liquid, force, electricity, heat, magnetism, light, and so on. Therefore, we can not only see the structure but also study the dynamic evolution of structures induced by these stimulations, namely materials science labs in

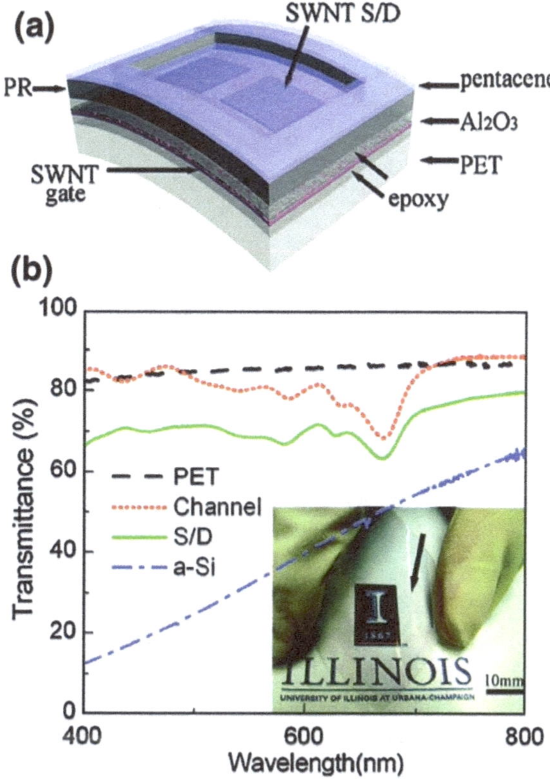

Fig. 1.11 Schematic geometry (**a**), optical transmission spectra (**b**) and optical image (*inset*) of organic transparent thin film transistors (TTFTs) based on SWCNTs on a plastic substrate. Reprinted with permission from Ref. [80]. Copyright 2006, American Institute of Physics

TEM. In this section, progresses of in situ TEM studies of CNTs will be introduced, including the growth mechanism, properties measurement, structural manipulation, and devices fabrication.

1.4.1 Studying CNTs Growth Mechanism by Using In Situ TEM

Traditionally, the growth mechanism was investigated by post-growth structural analysis, which lacked observation of the intermediate states during the CNTs growth. In situ TEM technique can correlate the growth conditions with product structures directly, and provide insight into the transitional states. Therefore, in situ TEM has been a powerful tool for growth mechanism studies, including environmental TEM and electron beam injection methods [81–87].

The environmental TEM method uses a modified TEM or a special sample holder to introduce gas into the TEM chamber and use resistance heating coils to provide a high temperature to mimic an environment for the growth of CNTs. During the nucleation and growth of CNTs, the state of the catalyst, the growth position, and carbon

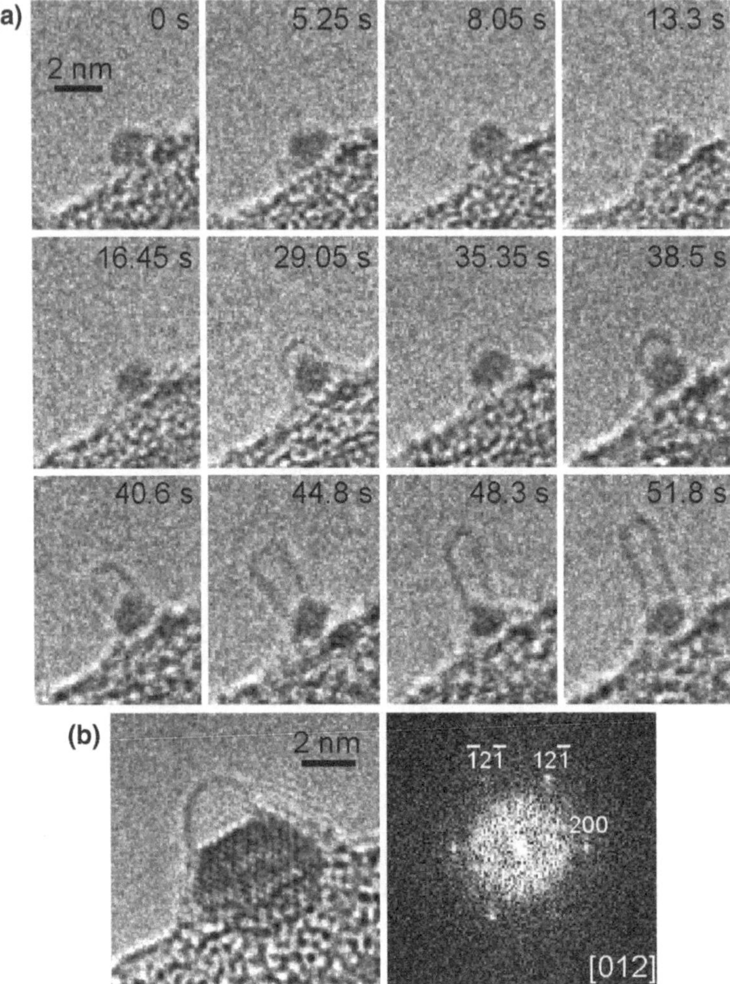

Fig. 1.12 In situ TEM observation of a SWCNT nucleation from a Fe catalyst (**a**), indicating the active catalyst is Fe_3C (**b**). Reprinted with the permission from Ref. [87]. Copyright 2008 American Chemical Society

atom diffusion route could be observed. Helveg et al. found that the catalyst for multi-walled CNTs (MWCNTs) are single crystalline Ni with fluctuating shapes, and carbon shells nucleate from the edges of the particle [81]. Yoshida et al. revealed that the catalysts for SWCNTs are single crystalline Fe_3C with clear crystalline lattices, and CNTs are formed by surface precipitation and cap lifting-up (Fig. 1.12) [87].

The electron beam injection method makes use of the interactions of electron beam and carbon atoms to inject the atoms into the inner spaces of a heated nanotube, which acts as a nano furnace. The injected atoms react with metal catalysts

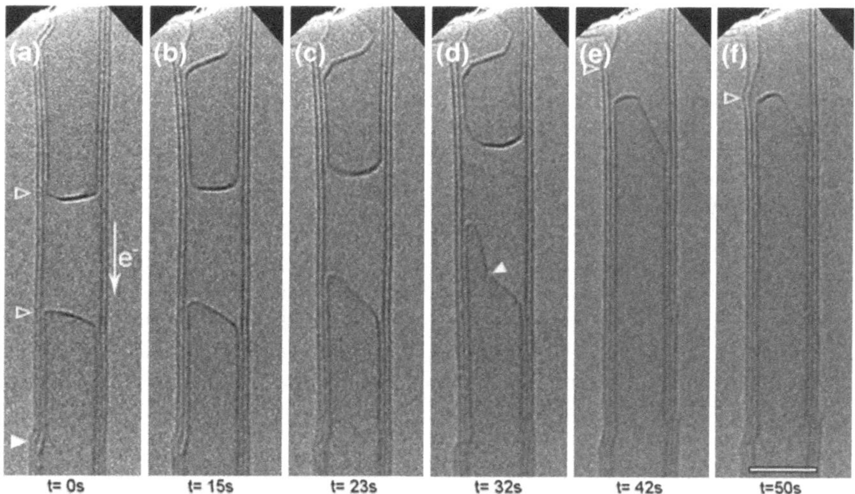

Fig. 1.13 *In situ* TEM investigation of the growth and shrinkage of a CNT by injecting carbon atoms at high temperature. Reprinted with the permission from Ref. [86]. Copyright 2008 American Chemical Society

filled in the nano furnace and forms carbon caps. Therefore, the state of the catalyst and the nucleation process of the nanotubes could be observed in situ [85]. For example, Jin et al. injected carbon atoms into a closed carbon cap and found that the nanotubes could be grown or shortened without the assistance of a catalyst and keeping the cap closed (Fig. 1.13) [86].

In situ TEM holds the possibility of direct correlation of the growth conditions and the atomic structures, and provides valuable guidelines for the controllable synthesis of CNTs. Up to now, most of the researches are concentrated on growth of CNTs catalyzed the Fe group metals, while many new catalysts were discovered, bringing new questions for the growth mechanism. Therefore, in this thesis we adopt in situ TEM technique to study the nucleation and growth mechanisms of CNTs from metal-free catalysts, and try to propose a general growth mechanism based on the comparative studies.

1.4.2 Properties Measurement by Using In Situ TEM

By applying stimuli such as electrical field or stress inside TEM, corresponding properties of CNTs could be measured and correlated with the structure. For example, Frank et al. reported quantized conductance by immersing CNTs into Hg [88]. Poncharal et al. measured the elastic properties of CNTs by applying an AC electrical field between a CNT and a counter electrode. The nanotube resonates when the applied frequency is close to its natural frequency (Fig. 1.14)

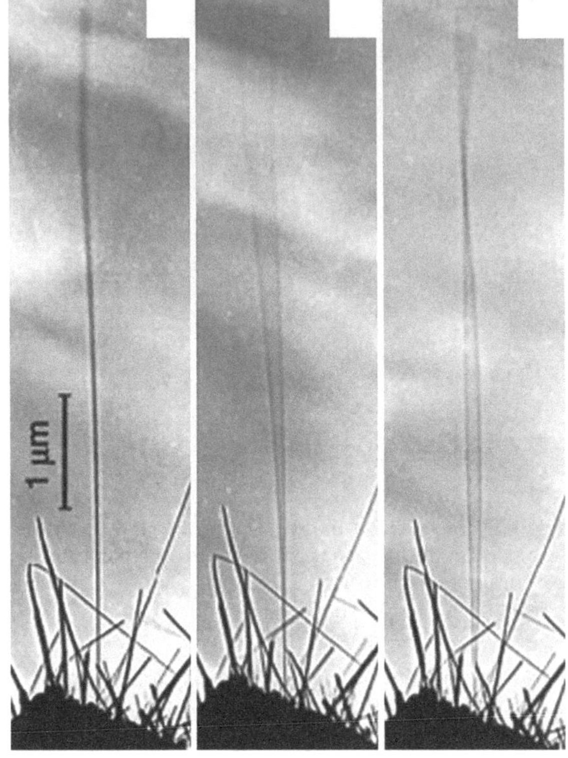

Fig. 1.14 Resonance of a CNT to the AC potential: (*left*) Initial nanotube, (*middle-right*) resonant vibration with potentials of 530 kHz and 1.03 MHz, respectively, from Ref. [89]. Reprint with permission from AAAS

[89]. Based on this technique, Jensen et al. designed an atomic mass sensor with atomic resolution [90].

1.4.3 Studying Structural Evolution of CNTs by Using In Situ TEM

In addition to the properties measurement, the applied stimuli could induce structural evolutions and may be used for structural manipulations. For example, Banhart et al. reported the bending, cutting, connection of CNTs, combining electron beam irradiation, and in situ heating techniques (Fig. 1.15) [91, 92]. Hetero-junctions between CNTs and metals were fabricated to measure the electrical and mechanical properties [93]. Suenaga and Iijima et al. investigated the formation and migration of point defects in CNTs using a focused electron beam [94]. By using the technique, Jin et al. fabricated carbon atomic chains and investigated the connection of small diameter CNTs [95, 96]. By passing a high current to induce a high temperature by Joule heating, Zettl et al. reported a layer by layer peeling of a MWCNT and Huang et al. reported the superplasticity of CNTs [97, 98].

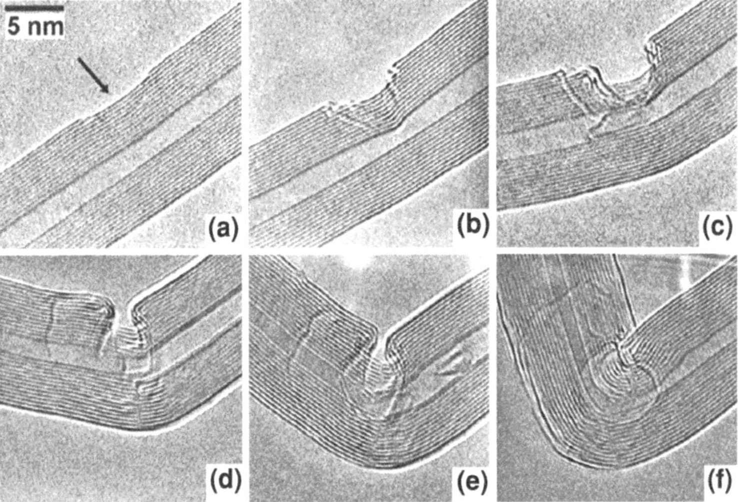

Fig. 1.15 Engineering of a CNT with a focused electron beam at high temperature. Reprinted with the permission from Ref. [92]. Copyright 2004 American Chemical Society

1.4.4 Fabrication of CNTs-Based Nanodevices by Using In Situ TEM

Based on the structural manipulation techniques, it is possible to fabricate CNT nanodevices. Combined with properties measurements, it is appealing to establish the fabrication-structure-properties relationships at atomic scale. For example, Cumings et al. removed the outer layers of a MWCNT, and fabricated a nanoscale piston by drawing and inserting the inner remaining nanotube, and measured the friction forces between the graphitic sheets [99]. Begtrup et al. made use of the electromigration effect to move the metal nanoparticles filled in the CNT and measured the change of resistance and proposed a new memory device (Fig. 1.16) [100]. The mass transport of nanoparticles outside nanotube was studied, and a mass conveyor was invented [101]. Bando group in National Institute for Materials Science (NIMS) invented the smallest thermometer by observing the volume changes of filled materials inside CNTs [102].

To summarize, the nano-lab inside TEM provides valuable opportunities to directly correlate the fabrication process, materials structures, and device properties at atomic resolution. By using the in situ TEM techniques, great progresses have been made in the CNT science from the understanding of growth mechanism, structural evolution and manipulation, to the devices fabrication and properties measurement. However, the nano-lab inside TEM is limited by the size of the pole piece and high vacuum; therefore it is necessary to combine in situ TEM with other methods such as ex situ experiments and theoretical simulations to achieve a comprehensive understanding.

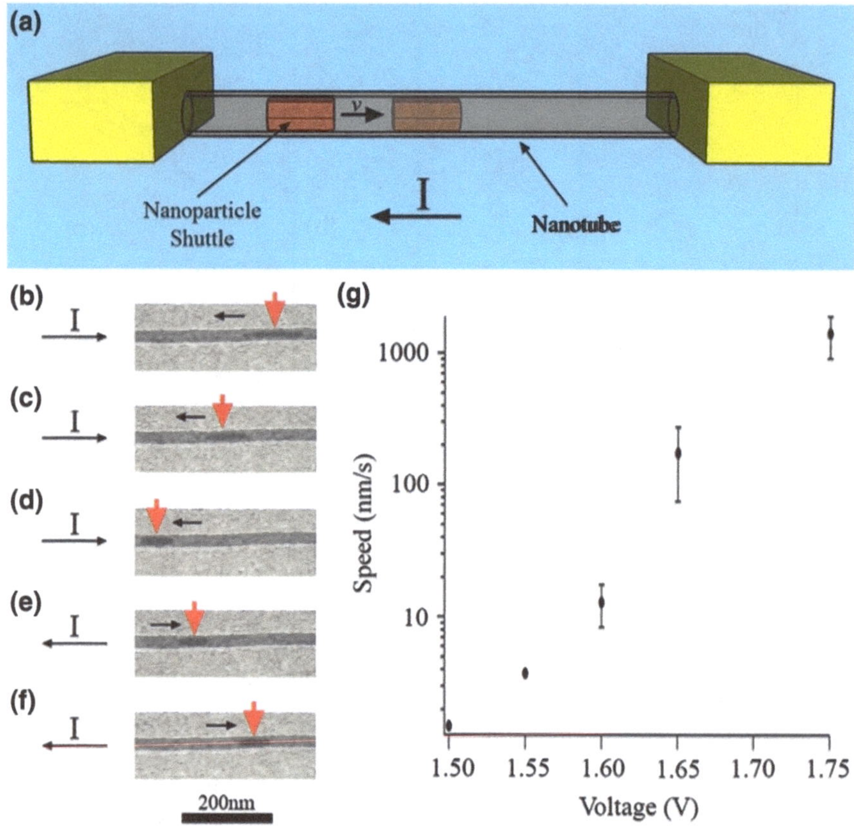

Fig. 1.16 TEM images and the bias-driven movement of a nanoshuttle memory device made of a metal-filled CNT in TEM. Reprinted with the permission from Ref. [100]. Copyright 2009 American Chemical Society

1.5 Motivations of the Thesis

During the past two decades, great advances have been made in CNT researches, however, there remain a lot of challenges as summarized as follows.

(a) The control of production scale, purity, and macroscale arrangement has reached a high level, however, selective synthesis of CNTs with certain conductivity and chirality is not effective.
(b) Researches on the traditional catalysts have been extensively conducted to understand their growth mechanisms. However, little is known about the growth mechanism of CNTs grown from newly discovered catalysts.
(c) Many prototype devices have been proposed and some of them are even commercially available. However, limited by the controllable synthesis, there is a lack of devices featured by the unique structure and properties of CNTs.

The research work in this thesis is motivated by above challenges. In situ TEM techniques are used to study the CNTs growth mechanisms by comparing the states of different catalysts. Based on the understandings, we suggest a new route for the controllable synthesis of CNTs. In addition, a CNT-clamped MAC is designed and fabricated to make use of the nanometer size and low electron scattering of CNTs.

References

1. Cheng H-M (2002) Synthesis, structure physical, properties and applications of carbon nanotubes. Chemical Industry Press, Beijing
2. Dresselhaus MS, Dresselhaus G, Eklund PC (1996) Science of fullerenes and carbon nanotubes. Elsevier, Amsterdam
3. Kroto HW, Heath JR, O'Brien SC, Curl RF, Smalley RE (1985) C60: Buckminsterfullerene. Nature 318:162–163
4. Iijima S (1991) Helical microtubules of graphitic carbon. Nature 354:56–58
5. Novoselov KS, Geim AK, Morozov SV, Jiang D, Zhang Y, Dubonos SV, Grigorieva IV, Firsov AA (2004) Electric field effect in atomically thin carbon films. Science 306:666–669
6. Endo M, Iijima S, Dresselhaus MS (1996) Carbon nanotubes. Elsevier, Amsterdam
7. Dresselhaus MS, Dresselhaus G, Avouris P (2001) Carbon nanotubes topics in applied physics. Springer, Berlin
8. Jorio A, Dresselhaus G, Dresselhaus MS (2008) Carbon nanotubes, advanced topics in the synthesis, structure properties and applications. Springer, Berlin
9. Harris PJF (2009) Carbon nanotube science synthesis, properties and applications. Cambridge University Press, Cambridge
10. Baughman RH, Zakhidov AA, de Heer WA (2002) Carbon nanotubes—the route toward applications. Science 297:787–792
11. Ajayan PM, Zhou OZ (2001) Applications of carbon nanotubes. Carbon Nanotubes 80:391–425
12. Endo M, Strano MS, Ajayan PM (2008) Potential applications of carbon nanotubes. Carbon Nanotubes 111:13–61
13. Saito R, Fujita M, Dresselhaus G, Dresselhaus MS (1992) Electronic structure of chiral graphene tubules. Appl Phys Lett 60:2204–2206
14. Iijima S, Ichihashi T (1993) Single-shell carbon nanotubes of 1-nm diameter. Nature 363:603–605
15. Bethune DS, Klang CH, de Vries MS, Gorman G, Savoy R, Vazquez J, Beyers R (1993) Cobalt-catalysed growth of carbon nanotubes with single-atomic-layer walls. Nature 363:605–607
16. Thess A, Lee R, Nikolaev P, Dai H, Petit P, Robert J, Xu C, Lee YH, Kim SG, Rinzler AG, Colbert DT, Scuseria GE, Tománek D, Fischer JE, Smalley RE (1996) Crystalline ropes of metallic carbon nanotubes. Science 273:483–487
17. Cheng HM, Li F, Su G, Pan HY, He LL, Sun X, Dresselhaus MS (1998) Large-scale and low-cost synthesis of single-walled carbon nanotubes by the catalytic pyrolysis of hydrocarbons. Appl Phys Lett 72:3282–3284
18. Kong J, Soh HT, Cassell AM, Quate CF, Dai H (1998) Synthesis of individual single-walled carbon nanotubes on patterned silicon wafers. Nature 395:878–881
19. Kong J, Cassell AM, Dai HJ (1998) Chemical vapor deposition of methane for single-walled carbon nanotubes. Chem Phys Lett 292:567–574
20. Zhang YG, Chang AL, Cao J, Wang Q, Kim W, Li YM, Morris N, Yenilmez E, Kong J, Dai HJ (2001) Electric-field-directed growth of aligned single-walled carbon nanotubes. Appl Phys Lett 79:3155–3157

21. Huang SM, Maynor B, Cai XY, Liu J (2003) Ultralong, well-aligned single-walled carbon nanotube architectures on surfaces. Adv Mater 15:1651
22. Ismach A, Segev L, Wachtel E, Joselevich E (2004) Atomic-step-templated formation of single wall carbon nanotube patterns. Angewandte Chemie-Int Ed 43:6140–6143
23. Liu B, Ren W, Liu C, Sun C-H, Gao L, Li S, Jiang C, Cheng H-M (2009) Growth velocity and direct length-sorted growth of short single-walled carbon nanotubes by a metal-catalyst-free chemical vapor deposition process. ACS Nano 3:3421–3430
24. Hata K, Futaba DN, Mizuno K, Namai T, Yumura M, Iijima S (2004) Water-assisted highly efficient synthesis of impurity-free single-waited carbon nanotubes. Science 306:1362–1364
25. Hayamizu Y, Yamada T, Mizuno K, Davis RC, Futaba DN, Yumura M, Hata K (2008) Integrated three-dimensional microelectromechanical devices from processable carbon nanotube wafers. Nat Nanotechnol 3:289–294
26. Futaba DN, Hata K, Yamada T, Hiraoka T, Hayamizu Y, Kakudate Y, Tanaike O, Hatori H, Yumura M, Iijima S (2006) Shape-engineer able and highly densely packed single-walled carbon nanotubes and their application as super-capacitor electrodes. Nat Mater 5:987–994
27. Mizuno K, Ishii J, Kishida H, Hayamizu Y, Yasuda S, Futaba DN, Yumura M, Hata K (2009) A black body absorber from vertically aligned single-walled carbon nanotubes. Proc Natl Acad Sci USA 106:6044–6047
28. Ding L, Yuan DN, Liu J (2008) Growth of high-density parallel arrays of long single-walled carbon nanotubes on quartz substrates. J Am Chem Soc 130:5428
29. Bachilo SM, Balzano L, Herrera JE, Pompeo F, Resasco DE, Weisman RB (2003) Narrow (n, m)-distribution of single-walled carbon nanotubes grown using a solid supported catalyst. J Am Chem Soc 125:11186–11187
30. Kim W, Choi HC, Shim M, Li Y, Wang D, Dai H (2002) Synthesis of Ultralong and high percentage of semiconducting single-walled carbon nanotubes. Nano Lett 2:703–708
31. Li XL, Tu XM, Zaric S, Welsher K, Seo WS, Zhao W, Dai HJ (2007) Selective synthesis combined with chemical separation of single-walled carbon nanotubes for chirality selection. J Am Chem Soc 129:15770
32. Qu L, Du F, Dai L (2008) Preferential syntheses of semiconducting vertically aligned single-walled carbon nanotubes for direct use in FETs. Nano Lett 8:2682–2687
33. Chiang W-H, Mohan Sankaran R (2009) Linking catalyst composition to chirality distributions of as-grown single-walled carbon nanotubes by tuning NixFe1-x nanoparticles. Nat Mater 8:882–886
34. Ding L, Tselev A, Wang J, Yuan D, Chu H, McNicholas TP, Li Y, Liu J (2009) Selective growth of well-aligned semiconducting single-walled carbon nanotubes. Nano Lett 9:800–805
35. Harutyunyan AR, Chen G, Paronyan TM, Pigos EM, Kuznetsov OA, Hewaparakrama K, Kim SM, Zakharov D, Stach EA, Sumanasekera GU (2009) Preferential growth of single-walled carbon nanotubes with metallic conductivity. Science 326:116–120
36. Hong G, Zhang B, Peng BH, Zhang J, Choi WM, Choi JY, Kim JM, Liu ZF (2009) Direct growth of semiconducting single-walled carbon nanotube array. J Am Chem Soc 131:14642
37. Li X, Tu X, Zaric S, Welsher K, Seo WS, Zhao W, Dai H (2007) Selective synthesis combined with chemical separation of single-walled carbon nanotubes for chirality selection. J Am Chem Soc 129:15770–15771
38. Wagner RS, Ellis WC (1964) Vapor-liquid-solid mechanism of single crystal growth. Appl Phys Lett 4:89
39. Baker RTK, Barber MA, Harris PS, Feates FS, Waite RJ (1972) Nucleation and growth of carbon deposits from the nickel catalyzed decomposition of acetylene. J Catal 26:51–62
40. Tibbetts GG (1984) Why are carbon filaments tubular? J Cryst Growth 66:632–638
41. Li Y, Kim W, Zhang Y, Rolandi M, Wang D, Dai H (2001) Growth of single-walled carbon nanotubes from discrete catalytic nanoparticles of various sizes. J Phys Chem B 105:11424–11431
42. Li F (2001) Synthesis and physical properties of single-walled carbon nanotubes by catalytic decomposition of hydrocarbons. PhD thesis, Chinese Academy of Science, Shenyang
43. Ren W-C (2005) Controllable synthesis, growth mechanism and physical properties of carbon nanotubes. PhD thesis, Chinese Academy of Sciences, Shenyang

44. Ren WC, Li F, Cheng HM (2006) Evidence for, and an understanding of, the initial nucleation of carbon nanotubes produced by a floating catalyst method. J Phys Chem B 110:16941–16946
45. Gavillet J, Loiseau A, Journet C, Willaime F, Ducastelle F, Charlier JC (2001) Root-growth mechanism for single-wall carbon nanotubes. Phys Rev Lett 87:275504
46. Maiti A, Brabec CJ, Bernholc J (1997) Kinetics of metal-catalyzed growth of single-walled carbon nanotubes. Phys Rev B 55:R6097–R6100
47. Ding F, Bolton K, Rosén A (2004) Nucleation and growth of single-walled carbon nanotubes: a molecular dynamics study. J Phys Chem B 108:17369–17377
48. Ding F, Larsson P, Larsson JA, Ahuja R, Duan H, Rosén A, Bolton K (2007) The importance of strong carbon—metal adhesion for catalytic nucleation of single-walled carbon nanotubes. Nano Lett 8:463–468
49. Zhou W, Han Z, Wang J, Zhang Y, Jin Z, Sun X, Zhang Y, Yan C, Li Y (2006) Copper catalyzing growth of single-walled carbon nanotubes on substrates. Nano Lett 6:2987–2990
50. Ritschel M, Leonhardt A, Elefant D, Oswald S, Büchner B (2007) Rhenium-catalyzed growth carbon nanotubes. J Phys Chem C 111:8414–8417
51. Liu B, Ren W, Gao L, Li S, Liu Q, Jiang C, Cheng H-M (2008) Manganese-catalyzed surface growth of single-walled carbon nanotubes with high efficiency. J Phys Chem C 112:19231–19235
52. Takagi D, Kobayashi Y, Hibino H, Suzuki S, Homma Y (2008) Mechanism of gold-catalyzed carbon material growth. Nano Lett 8:832–835
53. Yuan D, Ding L, Chu H, Feng Y, McNicholas TP, Liu J (2008) Horizontally aligned single-walled carbon nanotube on quartz from a large variety of metal catalysts. Nano Lett 8:2576–2579
54. Zhang Y, Zhou W, Jin Z, Ding L, Zhang Z, Liang X, Li Y (2008) Direct growth of single-walled carbon nanotubes without metallic residues by using lead as a catalyst. Chem Mater 20:7521–7525
55. Takagi D, Hibino H, Suzuki S, Kobayashi Y, Homma Y (2007) Carbon nanotube growth from semiconductor nanoparticles. Nano Lett 7:2272–2275
56. Liu B, Ren W, Gao L, Li S, Pei S, Liu C, Jiang C, Cheng H-M (2009) Metal-catalyst-free growth of single-walled carbon nanotubes. J Am Chem Soc 131:2082–2083
57. Huang S, Cai Q, Chen J, Qian Y, Zhang L (2009) Metal-catalyst-free growth of single-walled carbon nanotubes on substrates. J Am Chem Soc 131:2094–2095
58. Steiner SA, Baumann TF, Bayer BC, Blume R, Worsley MA, MoberlyChan WJ, Shaw EL, Schlögl R, Hart AJ, Hofmann S, Wardle BL (2009) Nanoscale zirconia as a nonmetallic catalyst for Graphitization of carbon and growth of single- and multiwall carbon nanotubes. J Am Chem Soc 131:12144–12154
59. Takagi D, Kobayashi Y, Homma Y (2009) Carbon nanotube growth from diamond. J Am Chem Soc 131:6922–6923
60. Rao F, Li T, Wang Y (2009) Growth of "all-carbon" single-walled carbon nanotubes from diamonds and fullerenes. Carbon 47:3580–3584
61. Yao Y, Feng C, Zhang J, Liu Z (2009) "Cloning" of single-walled carbon nanotubes via open-end growth mechanism. Nano Lett 9:1673–1677
62. Liu H, Takagi D, Chiashi S, Homma Y (2010) The growth of single-walled carbon nanotubes on a silica substrate without using a metal catalyst. Carbon 48:114–122
63. Homma Y, Liu HP, Takagi D, Kobayashi Y (2009) Single-walled carbon nanotube growth with non-iron-group "catalysts" by chemical vapor deposition. Nano Res 2:793–799
64. Javey A, Guo J, Wang Q, Lundstrom M, Dai HJ (2003) Ballistic carbon nanotube field-effect transistors. Nature 424:654–657
65. Landauer R (1970) Electrical resistance of disordered one-dimensional lattices. Philos Mag 21:863
66. White CT, Todorov TN (2001) Quantum electronics—Nanotubes go ballistic. Nature 411:649–651
67. Mann D, Javey A, Kong J, Wang Q, Dai H (2003) Ballistic transport in metallic nanotubes with reliable pd ohmic contacts. Nano Lett 3:1541–1544

68. Poncharal P, Berger C, Yi Y, Wang ZL, de Heer WA (2002) Room temperature ballistic conduction in carbon nanotubes. J Phys Chem B 106:12104–12118
69. Postma HWC, Teepen T, Yao Z, Grifoni M, Dekker C (2001) Carbon nanotube single-electron transistors at room temperature. Science 293:76–79
70. Tang ZK, Zhang L, Wang N, Zhang XX, Wen GH, Li GD, Wang JN, Chan CT, Sheng P (2001) Superconductivity in 4 angstrom single-walled carbon nanotubes. Science 292:2462–2465
71. Tsukagoshi K, Alphenaar BW, Ago H (1999) Coherent transport of electron spin in a ferromagnetically contacted carbon nanotube. Nature 401:572–574
72. Martel R, Derycke V, Lavoie C, Appenzeller J, Chan KK, Tersoff J, Avouris P (2001) Ambipolar electrical transport in semiconducting single-wall carbon nanotubes. Phys Rev Lett 87:256805
73. Avouris P, Chen ZH, Perebeinos V (2007) Carbon-based electronics. Nat Nanotechnol 2:605–615
74. Martel R, Schmidt T, Shea HR, Hertel T, Avouris P (1998) Single- and multi-wall carbon nanotube field-effect transistors. Appl Phys Lett 73:2447–2449
75. Tans SJ, Verschueren ARM, Dekker C (1998) Room-temperature transistor based on a single carbon nanotube. Nature 393:49–52
76. Soh HT, Quate CF, Morpurgo AF, Marcus CM, Kong J, Dai H (1999) Integrated nanotube circuits: controlled growth and ohmic contacting of single-walled carbon nanotubes. Appl Phys Lett 75:627–629
77. Wind SJ, Appenzeller J, Martel R, Derycke V, Avouris P (2002) Vertical scaling of carbon nanotube field-effect transistors using top gate electrodes. Appl Phys Lett 80:3817–3819
78. Bockrath M, Hone J, Zettl A, McEuen PL, Rinzler AG, Smalley RE (2000) Chemical doping of individual semiconducting carbon-nanotube ropes. Phys Rev B 61:R10606
79. Derycke V, Martel R, Appenzeller J, Avouris P (2001) Carbon nanotube inter- and intramolecular logic gates. Nano Lett 1:453–456
80. Cao Q, Zhu ZT, Lemaitre MG, Xia MG, Shim M, Rogers JA (2006) Transparent flexible organic thin-film transistors that use printed single-walled carbon nanotube electrodes. Appl Phys Lett 88:113511
81. Helveg S, López-Cartes C, Sehested J, Hansen PL, Clausen BS, Rostrup-Nielsen JR, Abild-Pedersen F, Nørskov JK (2004) Atomic-scale imaging of carbon nanofibre growth. Nature 427:426–429
82. Sharma R, Iqbal Z (2004) In situ observations of carbon nanotube formation using environmental transmission electron microscopy. Appl Phys Lett 84:990–992
83. Lin M, Tan JPY, Boothroyd C, Loh KP, Tok ES, Foo YL (2006) Direct observation of single-walled carbon nanotube growth at the atomistic scale. Nano Lett 6:449–452
84. Hofmann S, Sharma R, Ducati C, Du G, Mattevi C, Cepek C, Cantoro M, Pisana S, Parvez A, Cervantes-Sodi F, Ferrari AC, Dunin-Borkowski R, Lizzit S, Petaccia L, Goldoni A, Robertson J (2007) In situ observations of catalyst dynamics during surface-bound carbon nanotube nucleation. Nano Lett 7:602–608
85. Rodríguez-Manzo JA, Terrones M, Terrones H, Kroto HW, Sun LT, Banhart F (2007) In situ nucleation of carbon nanotubes by the injection of carbon atoms into metal particles. Nat Nanotechnol 2:307–311
86. Jin C, Suenaga K, Iijima S (2008) How does a carbon nanotube grow? An in situ investigation on the cap evolution. ACS Nano 2:1275–1279
87. Yoshida H, Takeda S, Uchiyama T, Kohno H, Homma Y (2008) Atomic-scale in situ observation of carbon nanotube growth from solid state iron carbide nanoparticles. Nano Lett 8:2082–2086
88. Frank S, Poncharal P, Wang ZL, de Heer WA (1998) Carbon nanotube quantum resistors. Science 280:1744–1746
89. Poncharal P, Wang ZL, Ugarte D, de Heer WA (1999) Electrostatic deflections and electromechanical resonances of carbon nanotubes. Science 283:1513–1516

90. Jensen K, Kim K, Zettl A (2008) An atomic-resolution nanomechanical mass sensor. Nat Nanotechnol 3:533–537
91. Krasheninnikov AV, Banhart F (2007) Engineering of nanostructured carbon materials with electron or ion beams. Nat Mater 6:723–733
92. Li J, Banhart F (2004) The engineering of hot carbon nanotubes with a focused electron beam. Nano Lett 4:1143–1146
93. Rodríguez-Manzo JA, Banhart F, Terrones M, Terrones H, Grobert N, Ajayan PM, Sumpter BG, Meunier V, Wang M, Bando Y, Golberg D (2009) Heterojunctions between metals and carbon nanotubes as ultimate nanocontacts. Proc Natl Acad Sci USA 106:4591–4595
94. Suenaga K, Wakabayashi H, Koshino M, Sato Y, Urita K, Iijima S (2007) Imaging active topological defects in carbon nanotubes. Nat Nanotechnol 2:358–360
95. Jin C, Lan H, Peng L, Suenaga K, Iijima S (2009) Deriving carbon atomic chains from graphene. Phys Rev Lett 102:205501
96. Jin CH, Suenaga K, Iijima S (2008) Plumbing carbon nanotubes. Nat Nanotechnol 3:17–21
97. Cumings J, Collins PG, Zettl A (2000) Materials—peeling and sharpening multiwall nanotubes. Nature 406:586
98. Huang JY, Chen S, Wang ZQ, Kempa K, Wang YM, Jo SH, Chen G, Dresselhaus MS, Ren ZF (2006) Superplastic carbon nanotubes—conditions have been discovered that allow extensive deformation of rigid single-walled nanotubes. Nature 439:281
99. Cumings J, Zettl A (2000) Low-friction nanoscale linear bearing realized from multiwall carbon nanotubes. Science 289:602–604
100. Begtrup GE, Gannett W, Yuzvinsky TD, Crespi VH, Zettl A (2009) Nanoscale reversible mass transport for archival memory. Nano Lett 9:1835–1838
101. Regan BC, Aloni S, Ritchie RO, Dahmen U, Zettl A (2004) Carbon nanotubes as nanoscale mass conveyors. Nature 428:924–927
102. Gao Y, Bando Y (2002) Nanotechnology: Carbon nanothermometer containing gallium. Nature 415:599

Chapter 2
In Situ TEM Method and Materials

In 1959, Nobel laureate Richard Feynman gave a speech entitled "There's Plenty of Room at the Bottom". In his speech, he talked about the manipulation and control of materials on a small length scale, such as the fabrication of molecular machines. In 1982, Binnig et al. invented scanning tunneling microscope (STM) [1]. In 1990, Eigler et al. demonstrated the ability of arranging atoms and put 35 Xe atoms on the surface of Ni crystal patterned as "IBM" [2]. However, STM is based on the tunneling current and requires a conductive surface. Binning et al. invented atomic force microscopy (AFM) [3], which is based on the interacting forces between the scanning probe and sample surface. Recently, the resolution and manipulation precision of AFMs has reached atomic level [4]. What scanning probe microscopy (SPM) has in common is that the observations and manipulations are limited to the surface, lacking insights into the structural mechanisms. In addition, scanning is involved for both imaging and manipulation, which limits the efficiency. As already been introduced in Chap. 1, TEM is a powerful tool for the materials structural characterizations with atomic resolution. In recent years, it is possible to combine SPM with TEM to conduct high-resolution structural characterizations in three dimensions and precise manipulations of nanomaterials simultaneously [5].

In this thesis, in situ TEM-SPM techniques are used for the investigation of CNTs growth mechanism and fabrication of CNTs-based devices. In the following sections, the method will be introduced, including a short introduction of TEM with an emphasis on the interactions between electrons and materials, principles and functions of the TEM-SPM platform, first principles calculations, and design and fabrication of materials for the in situ experiments.

2.1 Interactions Between Electrons and Materials

TEM is a microscopic technique by using an electron beam transmitted through thin samples. According to the Abbe theory, TEM has the potential of extremely high subatomic resolution because of the short wavelength. The signals generated from the interactions between electrons and materials could be collected to get structural

D.-M. Tang, *In Situ Transmission Electron Microscopy Studies of Carbon Nanotube Nucleation Mechanism and Carbon Nanotube-Clamped Metal Atomic Chains*, Recognizing Outstanding Ph.D. Research, DOI: 10.1007/978-3-642-37259-9_2,

information, such as the morphology, crystalline phase structure, chemical bonding, composition, and so on. Since the invention of TEM by Knoll and Ruska in 1931, great progresses have been made in achieving higher resolution by increasing the accelerating voltage, improving the electromagnetic lens, and adopting the charge-coupled device (CCD) cameras for imagining. In addition, the 4D techniques developed by Zewail et al. enable the extremely fast processes in the range of femtoseconds [6]. In this thesis, the imaging function of TEM is used to record the structural evolutions; electron energy loss spectroscopy (EELS) and energy-dispersive X-ray spectroscopy (EDS) are applied to analyze the composition; and the interaction between electrons and materials are utilized for high precision machining.

2.1.1 *Elastic Interactions*

The interactions between electrons and materials include elastic and inelastic scattering. For the elastically scattered electrons, only the directions are changed, while for inelastically scattered electrons, both the direction and energy are changed. Elastically scattered electrons are involved in the formation of images, diffraction patterns, while inelastically scattered electrons contribute to the EELS and EDS.

The optical path diagrams for the different modes in TEM could be changed for different purposes. The electron beam emitted by the electron gun is transformed to be parallel by condenser lenses and reaches the surface of the sample. Most of the electrons transmit directly as the direct beam. The elastically scattered electrons form diffraction patterns on the focal plane, and form images on the image plane. By changing the current in the intermediate lens, a diffraction pattern or image is selected for further magnification and projected by the projection lens onto the CCD camera or negative films.

There are three types of imaging mechanisms and corresponding contrasts. The first type is the mass-thickness contrast, which is based on the differences of absorption of electrons in materials. The second type is diffraction contrast, which is based on the differences of diffraction of electron beams from different crystalline structures and different orientations. The third type is phase contrast for HRTEM images, which is based on the interferences of the direct beam and diffracted beams. The atomic structures such as the lattice constants and defects could be directly resolved in the HRTEM images. In this thesis, all the three contrasts are involved. For example, the mass-thickness contrast is used for the observation of morphologies and alignment of the probes. The diffraction contrast is used for the determination of phase structures of the filled metals. And the HRTEM images are used for the observation of atomic processes.

2.1.2 *Inelastic Interactions*

Besides elastic interactions, the high-energy electron beams also have strong inelastic interactions with the materials, including the secondary electrons, X-rays, and irradiation effects, and so on. The secondary electrons and X-rays are involved in the following one process. Because of inelastic scattering, part of the energy is transferred to the inner orbital electrons in the materials, resulting in the releasing of secondary electrons. Electrons in the higher energy orbital jump to the lower orbital and emit X-rays in the meantime. The excitation and emission are both related to the featured energy levels of elements and chemical bonding, therefore, the composition of the materials could be obtained by collecting the X-rays or measuring the energy loss of the transmitted electrons.

The above-mentioned elastic and inelastic interactions do not induce structural changes, while electron beam irradiation effects can alter the structure by transferring kinetic energy to the atoms [7–9]. Irradiation effects include four types: The first type is decomposition, which mainly happens in polymers. The long chain structure is destroyed by the ionization effect. The second type is heating effect, because of the interaction of incident electrons and phonons in the materials. The heating effect is strongly associated with the chemical bonding of the samples. For metals with high thermal conductivity, the heating effect is around several kelvins and negligible in normal cases. For ceramics and polymers with poor thermal conductivity, the temperature rise could be as much as 1000 K. In this thesis, because of the good thermal conductance of CNTs and filled metals, the heating effects are negligible. The third irradiation effect is knock-on effect, by which the atoms in the crystal lattices are knocked-off the equilibrium positions. This effect could be described by the following formula, $E_t = \left[\sqrt{100 + AE_d/5} - 10\right]/20$, where E_d is the energy of the incident beam, E_t is the transferred energy, and A is a constant associated with the element. For each element, there is threshold energy that the atom could be knocked-off. Generally, lighter elements have lower threshold energy. For example, carbon and iron atoms could be knocked out by an incident beam of 80 kV and 400 kV, respectively. In this thesis, the accelerating voltage is 200 kV or 300 kV, which can destroy the carbon shells and is not high enough to knock off the metal atoms, therefore realizing element-selective etching for the fabrication of CNT-clamped MACs. Also the technique is used for the injection of carbon atoms into the nano furnaces. Another electron irradiation effect is the surface sputtering. Because of the unsaturated chemical bonding on the surface, the threshold energy for displacing atoms is lower than that of bulk. Take Fe as an example, the energy barrier for bulk lattice and surface atoms are 16 eV and 4–8 eV, respectively. The energy transferred to Fe atoms from 200 kV and 300 kV electron beams are 9.4 eV and 15.3 eV, respectively, which is enough to knock off the surface atoms. In this thesis, the surface sputtering effects are involved in the thinning of exposed metal nanowires.

2.2 TEM-SPM In Situ Platform

In this thesis, a commercial TEM-SPM in situ platform from Nanofactory is used, including TEM-STM type for electrical measurements and TEM-AFM type for mechanical measurements. In the following parts, the configuration and functions of the platform will be introduced.

There are four main parts included in the TEM-SPM platform: a sample holder with a 3D piezo-motor, an electrical and mechanical testing system, and the control software. The signals of movements and bias are controlled by the software and sent through the control box and amplifier, and then to the probes inside the TEM. The measured currents and forces signals are collected inside the TEM, amplified and then sent back to the user interface on the computer. In the meantime, structural and compositional information is recorded the same as the normal TEM. Realtime structural evolutions are recorded by a CCD camera.

The difference in holder between this platform and normal TEM is the structure of the head. Instead of a hole for placing thin foils, a piezo tube connected ceramic ball and a hole are placed in the TEM-STM holder for the probe and the samples. For the TEM-AFM holder, a TEM cantilever is placed at the opposite side of the piezo tube. Therefore, by moving the probe against the cantilever, corresponding deflection and forces could be recorded. There is no difference in the overall outer shape and dimensions; therefore, the in situ holders are compatible with normal TEM. And tilting, selected area electron diffraction pattern (SAED), EDS, and EELS could be operated as normal.

A critical part of the TEM-SPM platform is the probe, because the size of the probe determines the length scale of the investigation system, and the contact between the probe and the sample determines contact resistance and reliability. The probes in this thesis are produced using a traditional electrochemical etching method [10]. The optimized parameters are as follows: DC bias 5 V, 1:1 hydrochloric acid (HCl), immersion length in electrolyte: 4 mm, counter electrode: gold wire, and etching time: 5 min. During the etching, the following reactions take place [11]:

$$\mathrm{Au} + 4\mathrm{Cl}^- \rightarrow \mathrm{AuCl}_4^- + 3e^-,\ E_0 = 1.002\,\mathrm{V};$$
$$\mathrm{Au} + 2\mathrm{Cl}^- \rightarrow \mathrm{AuCl}_2^- + e^-,\ E_0 = 1.154\,\mathrm{V};$$
$$\mathrm{AuCl}_2^- + 2\mathrm{Cl}^- \rightarrow \mathrm{AuCl}_4^- + 2e^-,\ E_0 = 0.926\,\mathrm{V}.$$

The etching speed at the liquid–air interface is fastest because the immersed part is coated with a viscous layer from the reactions. The wire will be disconnected by the weight of the immersed part when it is etched thin enough. Then the etched tips are cleaned by flushing water to stop further etching. The tips produced under optimized conditions could be as sharp as 50 nm in radius, as demonstrated by the SEM image (Fig. 2.1).

There are two modes for the movement control of the scanning probe. One is the coarse mode driven by vibration and inertial movements, with the movement range as large as 1 mm, so as to align the probe with the sample roughly. The other is the fine mode based on the piezoelectric principle, with very high precision up to 0.2 Å in the up/down and left/right directions and 0.025 Å in the backward/forward direction.

When the probe and sample are aligned precisely, various functions could be realized by applying corresponding stimuli. For example, the basic functions of electrical and mechanical measurements could be carried out by applying bias and moving against the AFM cantilever, respectively. In addition, the combination of Joule heating-induced high temperature, electromigration, electric field, and

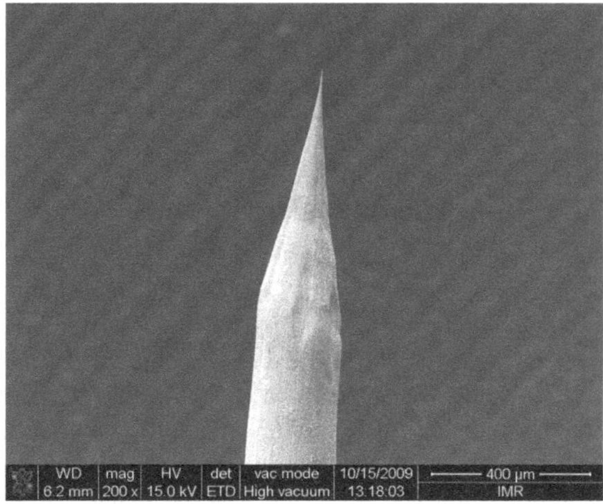

Fig. 2.1 SEM image of the tip of a Au probe

electron beam irradiation could realize more complicated functions. In the following parts, some examples will be presented to introduce the mains functions of the TEM-SPM platform.

First, it is possible to measure the intrinsic electrical properties of nanostructures and associate them with the HRTEM images. The applied bias range is $\pm$ 10 V ($\pm$ 140 V for field emission option), and the measurement noise is around 1 nA, depending on the circumstances. Jin et al. investigated the field emission of single CNTs by applying a high voltage between the probe and the CNT with a narrow gap [12]. It was found that along with the field emission, structural evolutions took place for the cap of the CNT, which was attributed as the degradation of the emitted currents. By passing a current, high temperature up to 2000 K could be obtained by Joule heating effect, and Huang et al. used this technique to study the deformation of CNTs at high temperature [13]. Superplasticity up to 280 % elongation was observed for SWCNTs.

Another basic function of the TEM-SPM platform is mechanical measurements. The noise of the measured force is at the nanoNewton level. Therefore, the force–displacement relationship of extremely small nanostructures could be tested. Golberg group in NIMS did systematic researches on the mechanical properties of 1D materials. For examples, it was found that under small bending angles, boron nitride (BN) nanotubes could be bent repeatedly and elastically, with the elastic modulus of 0.5–0.6 TPa, while permanent deformation occurred when the bending angle was larger than 115°, following the kink mechanism with buckling between the nanotube walls [14].

Besides the basic functions, various manipulations could be realized by the combination of electron beam irradiation, current, force, and so on. Svensson et al. used single CNTs as nanoscale pipettes to transfer nanoparticles by reversing the current direction and correspondingly the interaction forces [15]. Cumings et al. peeled off the outer layers of a MWCNT with Joule heating-induced high temperature and the inner parts of the CNT could be retracted and inserted like a nanoscale piston [16]. Sun et al. combined the electron beam irradiation and Joule heating to shrink a nanotube and applied a high pressure on the filled metals to observe the structural evolutions [17].

2.3 First Principles Calculations

The advantages of in situ TEM are the real-time observations of the structural evolutions and correlations with the properties measurements. However, critical questions such as the driving force could not be answered by using in situ TEM alone. In addition, the recorded TEM images are two-dimensional projection of the three-dimensional structure. Therefore, it is difficult to capture the structural evolutions comprehensively. Another shortcoming of in situ TEM is the limited temporal resolution. To acquire a TEM image, it takes at least 0.05 s, which is not short enough to record the fast processes. Therefore, in this thesis, theoretical calculations are used to complement the in situ TEM study. Proper structural models are built according to the TEM observations. And by comparing the total energy, stability of possible structures and evolution paths, the possible structural mechanisms could be revealed. In addition, the electric properties could be predicted and compared with the experimental results.

In this thesis, first principles calculations based on density functional theory (DFT) are used, and generalized gradient approximation (GGA) is adopted to describe the electronic exchange–correlation interactions [18]. All the calculations are conducted with the full-potential projector augmented-wave (PAW) method [19] in Vienna ab initio simulation package (VASP) [20]. Take the formation process of metal atomic chains as an example (110) (111) and (001) oriented nanowires were constructed for BCC structured Fe, FCC structured Pt and HCP structured Co, respectively. Tensile elongation with a step of 0.4 Å was applied on the atoms at the end surfaces. After each step, full structural relaxation was carried out. The total energy was calculated after the formation of single atomic chains. After that the charge density distributions were calculated to predict the nature of the chemical bonding and spin-resolved density states were computed to predict the electronic properties.

2.4 Design and Fabrication of Samples for In Situ TEM

The samples for in situ TEM are not only the objects of investigations but also important functional components. There are certain restrictions on samples in TEM, such as thickness and stability under high vacuum. Therefore, it is necessary to design both the structure and fabrication methods. In this thesis, most of the experiments are based on filled CNTs as summarized in Table 2.1, which were produced by CVD or AAO template methods.

Table 2.1 Summary of the samples in this work

Sample	Fabrication method	Structural features	Functions	Subject
Fe_2O_3-filled CNT	AAO template	Metal catalyst filled in CNT	Growth furnace, metallic catalyst	CNTs growth mechanism
SiO_x-filled CNT	AAO template	Metal free catalyst in CNT	Growth furnace, non-metallic catalyst	
Fe-filled CNT	FCCVD	Fe-filled CNT	CNT sheath during fabrication; Mechanical clamp during tension; Electrical leads during IV test	CNT-clamped MAC
Fe–Ni-filled CNT	FCCVD	Fe alloy-filled CNT		
Pt-filled CNT	AAO template	Non-traditional metal for CNT growth		

2.4.1 CVD Method

Fe-filled CNTs were produced by using acetylene (C_2H_2) as the carbon source, ferrocene ($Fe(C_5H_5)_2$) as the catalyst precursor at 860 °C. The CNTs are highly aligned with high purity, as revealed by the SEM image (Fig. 2.2a). TEM image (Fig. 2.2b) demonstrates that the CNTs are uniform in diameter and with high percentage of Fe filling. The filled Fe nanowires are about 200–300 nm in length. HRTEM image (Fig. 2.2c) shows the high crystallinity with clear lattices corresponding to body-cubic centered (BCC) structured Fe.

Chlorine was found to be able to etch carbon during the growth of CNTs by Lv et al. [21], and therefore high percentage metal-filled CNTs could be obtained. With this technique, Fe alloys-filled CNTs were grown by using a mixture of ferrocene, nickelocene, and cobaltocene as the catalyst precursor. The optimized growth conditions are: the solution of the precursors in trichlorotoluene with the concentration of 10 % is injected into the growth furnace with a speed of 0.15 μl/min.The growth temperature is 860 °C. The flow rates for Ar and H_2 are 1500 and 300 ml/min, respectively.

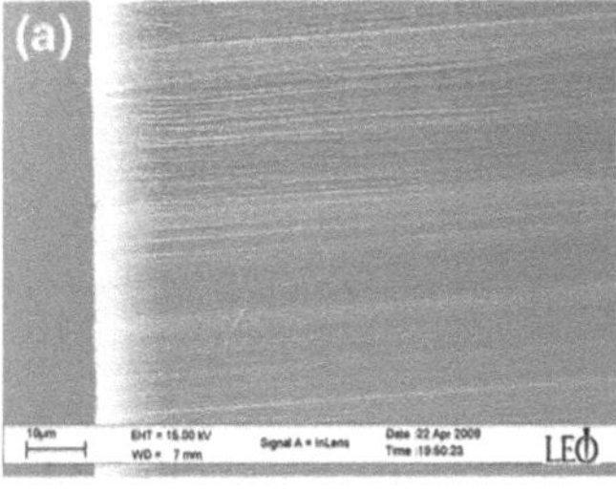

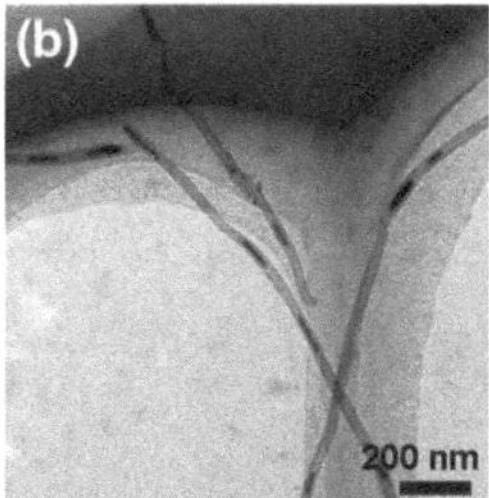

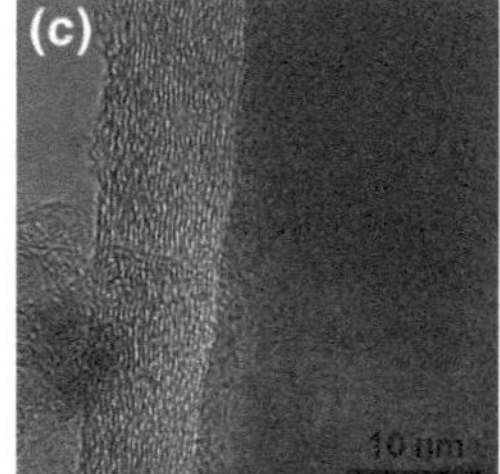

Fig. 2.2 Structure of the Fe-filled CNTs by FCCVD method using acetylene as carbon source. **a** SEM image. **b–c** Low-magnification and high-resolution TEM images

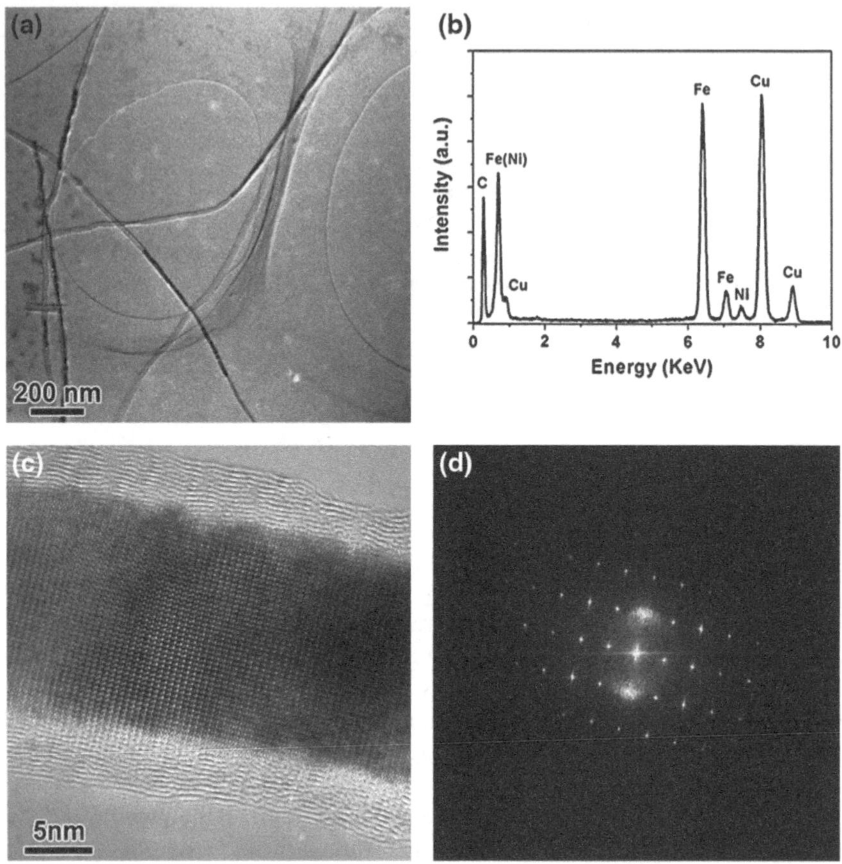

Fig. 2.3 Structure of the Fe–Ni alloy-filled CNTs by FCCVD method using chloride-included carbon source: **a** TEM image, **b** EDS spectrum, **c** High resolution TEM image, **d** FFT pattern of (**c**) [22]

Figure 2.3 shows the characterization results of the Fe–Ni alloy-filled CNTs. Compared with previous results of Fe-filled CNTs, the filling ratio and length are both increased. Fe, Ni, C, and Cu signals are detected by EDS, where Cu comes from the TEM grid. Because of the etching of chlorine, the walls of the CNTs are much thinner (Fig. 2.3c). Both the HRTEM image and FFT pattern (Fig. 2.3d) confirm the single crystalline nature of the filled alloy.

2.4.2 AAO Template Method

By using CVD method in the previous section, the filling of CNTs is limited to the catalyst metals, typically Fe-group metals and alloys. While various metals could be filled into CNTs by using the AAO template method, which will be

Table 2.2 Experimental parameters of SiO_x, Fe_2O_3, and Pt-filled CNTs by using AAO template method

Material	Catalog	Step one	Step two	Step three
SiO_x	Non-metal oxide	Production of AAO template	CNTs deposition	SiO_x deposition
Fe_2O_3	Metal oxide		CNTs deposition	$Fe(NO_3)_3$ decomposition
Pt	Noble metal		Pt nanowire deposition	CNT deposition

introduced in this section [23]. Pt, Fe_2O_3, and SiO_x-filled CNTs were produced by the method, as summarized in Table 2.2.

The first step was the fabrication of AAO template, with the following procedure:

1. Initial electrochemical oxidation of Al foils (99.99 %) at 20 V DC in 10 %wt H_2SO_4, at 10 °C for 2 h.
2. Remove the initial oxidation layer in H_3PO_4 at 60 °C to get an ordered pattern.
3. Secondary electrochemical oxidation with the same parameters for 4 h to get AAO template with one end open.
4. Open the other end by further anodic oxidation in a mixed electrolyte of $HClO_4$ and acetone (2:1) at 25 V for 1–3 min.

The second step for the preparation of filled CNTs by using AAO template method was the deposition of CNTs in the pores of the templates, with the following procedure:

1. Put the AAO template in the middle of a vertical CVD furnace; increase the temperature to 650 °C with a speed of 10 °C/min under N_2 flow (100 sccm).
2. Deposition of carbon at 650 °C by using C_2H_2 as the carbon source (10 sccm) and Ar as the carrier gas (100 sccm).
3. Annealing at 750 °C in N_2 for 3 h to enhance the crystalline order of the deposited carbon layers.
4. Remove the AAO template in NaOH solution (5 M) to get free standing CNTs.

The third step was to use a CVD method for the filling of SiO_x in CNTs with the following procedure:

1. Put the AAO template in the middle of a vertical CVD furnace; increase the temperature to 500 °C with a speed of 10 °C/min under N_2 flow (100 sccm).
2. Deposition of SiO_x at 500 °C, by using SiH_4 as the silicon source (10 sccm) and Ar as the carrier gas (40 sccm). During the deposition, trace amount of O_2 is introduced into the furnace to get SiO_x.

Figure 2.4 shows the TEM image of the SiO_x-filled CNTs. The CNTs fabricated from AAO templates are highly uniform in diameter and are filled with SiO_x particles. The CNTs are poor crystalline as revealed by the curved and non-continuous lattices. The SiO_x particles are 5–10 nm in diameter. HRTEM image reveals that the particles are amorphous and free of carbon coating.

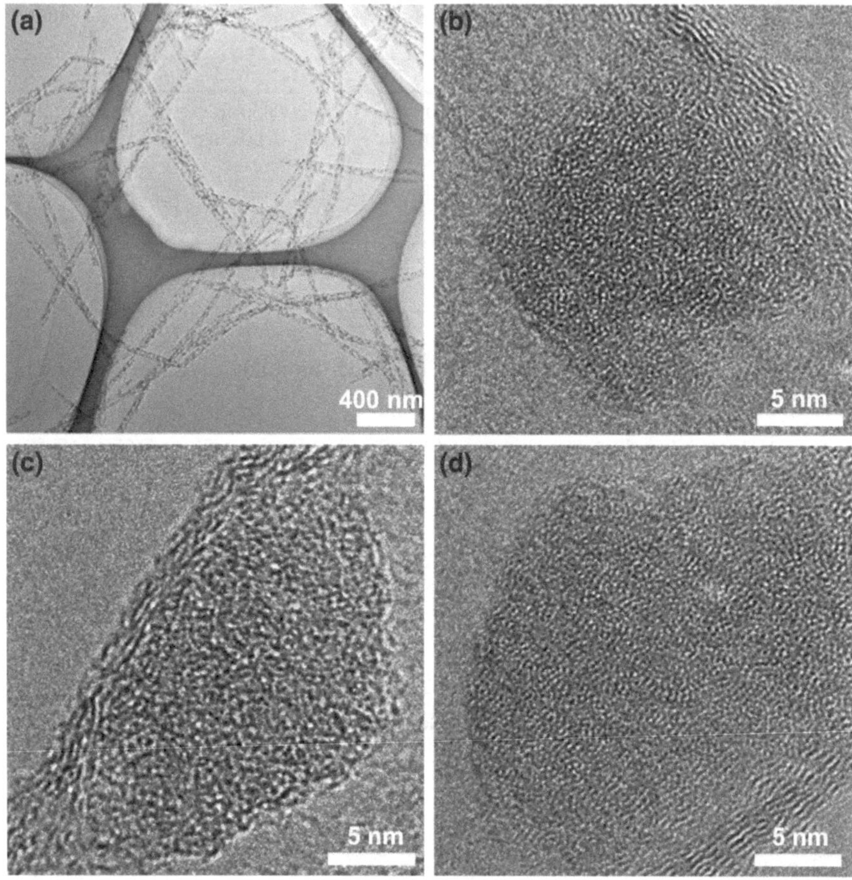

Fig. 2.4 Structural characterizations of the SiO_x-filled CNTs by using AAO template method: **a** TEM image, **b–d** High-resolution TEM images. Reprinted with the permission from Ref. [24]. Copyright 2011, American Chemical Society

The filling of Fe_2O_3 into CNTs was realized by a decomposition method, with the following procedure:

1. Immerse AAO template in the ethanol solution of $Fe(NO_3)_3$ at 80 °C for 30 min to get the precursor of Fe_2O_3 and then washed by deionized water.
2. Decomposition of the $Fe(NO_3)_3$ in AAO template at 350 °C in N_2 for 6 h.

Similar to the SiO_x-filled CNTs, the CNTs are uniform in diameter and the filling rate is very high. But the filled Fe_2O_3 nanoparticles are single crystalline, which is different from SiO_x. The diameters of the CNT and Fe_2O_3 particles are 50–60 nm and 5–10 nm, respectively.

Pt was filled into CNTs following another route by electrochemical deposition, with the following procedure:

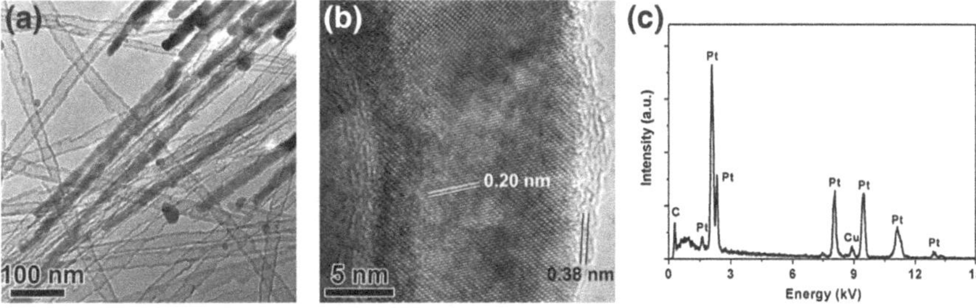

Fig. 2.5 Structure of the Pt-filled CNTs by AAO template method: **a** TEM image, **b** High-resolution TEM image, **c** EDS spectrum [22]

1. A layer of silver is deposited on one side of the AAO template as the electrode for further electrochemical deposition.
2. Electrochemical deposition of Pt using a three electrode system at −0.5 V for 1 h in the electrolyte of a mixture of H_2PtCl_6 (1.0 g/L) and H_3BO_3 (25 g/L).

Figure 2.5 shows the structure of Pt-filled CNTs. The CNTs are uniform in diameter and length. About 40 % of the CNTs are filled continuously. HRTEM image shows that the CNTs have very thin walls with only 3–5 layers. The filled Pt nanoparticles are single crystalline with lattices corresponding to Pt (111) as clearly identified. The composition was checked by EDS, and strong signals from Pt, C from Pt-filled CNTs, and Cu from TEM grids were detected.

2.5 Summary

Nano-lab inside TEM is an important research topic at the frontiers of nanoscience and nanotechnology, because of its atomic resolution, rich functions for measurement, and manipulations. In this chapter, the principle and method of in situ TEM-SPM technique is introduced, including the interactions between electrons and materials, functions of TEM-SPM in situ holders, first principles calculations, and fabrication of designated samples. The in situ platform combines the normal structural characterizations of TEM, materials processing functions by electron irradiation, various stimuli by the in situ holders, and theoretical calculations to provide in-depth understanding of the structural mechanism. The design and fabrication of suitable samples with specified structures and functions is the basis of realizing the functions. It is important to choose the right route for preparing samples according to the purpose of research, and then to select the right tools including experimental and also theoretical models to get a comprehensive understanding.

In Chap. 3, catalysts-filled CNTs are designed for the investigation of the CNTs growth mechanism. AAO template and CVD method are chosen for the filling of different kinds of catalyst. With Joule heating-induced high temperature and electron beam irradiation-induced carbon atoms injection, the growth of CNTs could be observed in situ. To realize the in situ investigations of the all-carbon growth of CNTs, carbon micro coils are designed, and high current-induced electromigration and Joule heating-induced high temperature are used to provide driving force.

In Chap. 4, metal-filled CNTs are designed for the fabrication of CNT-clamped MACs, and produced by using floating CVD and AAO template methods, depending on the chemical properties of the metals. Then element-selective electron beam irradiation is used for the peeling off of the carbon shells and surface sputtering is used to etch the metal nanowires. After that, tension is applied to elongate the nanowire and produce MACs. HRTEM and first principles calculations are conducted to study the formation process and the electrical properties of the fabricated CNT-clamped MACs are tested by using the TEM-STM in situ holder.

References

1. Binnig G, Rohrer H, Gerber C, Weibel E (1982) Tunneling through a controllable vacuum gap. Appl Phys Lett 40:178–180
2. Eigler DM, Schweizer EK (1990) Positioning single atoms with a scanning tunnelling microscope. Nature 344:524–526
3. Binnig G, Quate CF, Gerber C (1986) Atomic force microscope. Phys Rev Lett 56:930
4. Custance O, Perez R, Morita S (2009) Atomic force microscopy as a tool for atom manipulation. Nat Nano 4:803–810
5. Banhart F (2008) In-situ electron microscopy at high resolution. World Scientific, Singapore
6. Zewail AH (2010) Four-dimensional electron microscopy. Science 328:187–193
7. Egerton RF, Li P, Malac M (2004) Radiation damage in the TEM and SEM. Micron 35:399–409
8. Egerton RF, Wang F, Crozier PA (2006) Beam-induced damage to thin specimens in an intense electron probe. Microsc Microanal 12:65–71
9. Williams DB, Carter CB (2009) Transmission electron microscopy: a textbook for materials science. Springer, New York
10. Eligal L, Culfaz F, McCaughan V, Cade NI, Richards D (2009) Etching gold tips suitable for tip-enhanced near-field optical microscopy. Rev Sci Instrum 80:033701
11. Williams C, Roy D (2008) Fabrication of gold tips suitable for tip-enhanced Raman spectroscopy. J Vac Sci Technol, B 26:1761–1764
12. Jin CH, Wang JY, Wang MS, Su J, Peng LM (2005) In-situ studies of electron field emission of single carbon nanotubes inside the TEM. Carbon 43:1026–1031
13. Huang JY, Chen S, Wang ZQ, Kempa K, Wang YM, Jo SH, Chen G, Dresselhaus MS, Ren ZF (2006) Superplastic carbon nanotubes—Conditions have been discovered that allow extensive deformation of rigid single-walled nanotubes. Nature 439:281
14. Golberg D, Costa PMFJ, Lourie O, Mitome M, Bai X, Kurashima K, Zhi C, Tang C, Bando Y (2007) Direct force measurements and kinking under elastic deformation of individual multiwalled boron nitride nanotubes. Nano Lett 7:2146–2151
15. Svensson K, Olin H, Olsson E (2004) Nanopipettes for metal transport. Phys Rev Lett 93:145901
16. Cumings J, Zettl A (2000) Low-friction nanoscale linear bearing realized from multiwall carbon nanotubes. Science 289:602–604
17. Sun L, Banhart F, Krasheninnikov AV, Rodríguez-Manzo JA, Terrones M, Ajayan PM (2006) Carbon nanotubes as high-pressure cylinders and nanoextruders. Science 312:1199–1202
18. Wang Y, Perdew JP (1991) Correlation hole of the spin-polarized electron gas, with exact small-wave-vector and high-density scaling. Phys Rev B 44:13298

19. Blöchl PE (1994) Projector augmented-wave method. Phys Rev B 50:17953
20. Kresse G, Furthmüllerr J (1996) Efficient iterative schemes for ab initio total-energy calculations using a plane-wave basis set. Phys Rev B 54:11169
21. Lv RT, Kang FY, Wang WX, Wei JQ, Gu JL, Wang KL, Wu DH (2007) Effect of using chlorine-containing precursors in the synthesis of FeNi-filled carbon nanotubes. Carbon 45:1433–1438
22. Tang D-M, Yin L-C, Li F, Liu C, Yu W-J, Hou P-X, Wu B, Lee Y-H, Ma X-L, Cheng H-M (2010) Carbon nanotube-clamped metal atomic chain. Proc Natl Acad Sci USA 107:9055–9059
23. Kyotani T, Pradhan BK, Tomita A (1999) Synthesis of carbon nanotube composites in nanochannels of an anodic aluminum oxide film. Bull Chem Soc Jpn 72:1957–1970
24. Liu B, Tang D-M, Sun C, Liu C, Ren W, Li F, Yu W-J, Yin L-C, Zhang L, Jiang C, Cheng H-M (2011) Importance of oxygen in the metal-free catalytic growth of single-walled carbon nanotubes from SiOx by a vapor–solid–solid mechanism. J Am Chem Soc 133:197–199

Chapter 3
Studying Nucleation Mechanism of Carbon Nanotubes by Using In Situ TEM

The electrical and optical properties of CNTs are strongly correlated with the structures. Therefore, one of the most important topics for CNTs research is the structure-controllable synthesis. The key is the understanding of the nucleation and growth mechanism. In this chapter, the nucleation mechanism is investigated by using the in situ TEM method. First, the research background and progresses of CNTs growth mechanisms will be introduced. After that, the experimental procedures will be described. Then, the results of studying nucleation of CNTs on metal-free SiO_X catalyst using in situ TEM will be presented. Based on the observations and discussions, a general growth mechanism and a new method for structure-controllable synthesis are proposed. Then the growth of CNTs from all-carbon seeds, and heteroepitaxial growth from BN platelets will be demonstrated, followed by a summary.

3.1 Research Background

The progresses of CNTs growth have proved that the key to control the structure of CNTs is the catalyst. Therefore, the key to understand the growth mechanism is the role of catalyst. In this section, the growth mechanism will be presented for Fe group metals, new metallic catalysts, and metal-free catalysts.

3.1.1 CNTs Growth Mechanism on Traditional Fe Group Metals

Traditionally, the catalysts for CNTs growth are Fe group metals and alloys. Hydrocarbon gases can be catalytically decomposed by these metals and they can form stable or meta-stable carbide phases. The growth processes of CNTs by using Fe group metals as catalysts have been investigated intensively [1–7]. One of the

D.-M. Tang, *In Situ Transmission Electron Microscopy Studies of Carbon Nanotube Nucleation Mechanism and Carbon Nanotube-Clamped Metal Atomic Chains*, Recognizing Outstanding Ph.D. Research, DOI: 10.1007/978-3-642-37259-9_3,

widely accepted growth mechanisms is the VLS mechanism: carbon precursors are absorbed and catalytically decomposed on the surface of catalyst. Then carbon atoms are dissolved into the particle and form a liquid carbide phase. As the concentration of carbon increases to oversaturation, CNTs are precipitated.

Fe group catalysts have many advantages. The growth rate is very high because of the catalytic decomposition abilities and the fast diffusion in the liquid or quasi-liquid particles. High-efficiency production has been reported with the growth rate as high as 10 μm/s [8, 9]. However, due to the low melting points, the catalysts are liquid with fluctuating structures at the growth temperatures. Therefore, it is difficult to precisely control the structure of the catalyst particle and the grown CNTs [10–12].

3.1.2 CNTs Growth Mechanism on Newly Developed Metal Catalysts

As introduced before, the key to control CNTs structure is the catalysts. Therefore, non-Fe group metals used for the growth of CNTs have been investigated so as to understand the growth mechanism comprehensively, and possibly alternative approaches could be found to control CNTs structures. Many catalysts have been reported, such as Pd [13], Pt [13], Au [13–15], Ag [13], Cu [16], Re [17], Mn [18], Pb [19], Al [20], Mo [20], Cr [20], Mg [20], and so on. Most of the newly developed catalysts do not have the ability to catalytically decompose of hydrocarbon gases, and most of them cannot form carbides. Therefore, there must be new growth mechanisms for the CNTs from the new metallic catalysts. Homma et al. proposed a modified VLS growth mechanism involving a size-dependent solubility [15, 21]. In this growth model, the solubility of carbon in nanosized metal particles increases dramatically. Therefore, CNTs could be grown by the dissolution and precipitation approach. Different from the traditional VLS mechanism, a carbide phase is not involved in this modified mechanism.

3.1.3 CNTs Growth Mechanism on Recently Discovered Metal-Free Catalysts

Besides metal catalysts, it was found that metal-free catalysts could also grow CNTs and have many merits. For example, the absence of metallic contaminations makes it possible to study the intrinsic electrical and magnetic properties of CNTs. In semiconductor industry, possible short circuit caused by metals could be avoided. And metal-free CNTs can avoid the poisonous effects of metals in biology applications. In addition, metal-free catalyst is also helpful to enrich the understanding of the growth mechanisms.

In 2007, Homma group in Tokyo University of Sciences reported the growth of CNTs from semiconductors such as Ge, Si, and SiC by using molecular beam epitaxy (MBE) techniques [22]. In 2009, our group [23] and Huang group [24] from Wenzhou University reported the high-efficiency growth of CNTs from silicon oxide. After that, CNTs grown with metal-free catalysts attracted widespread interest [25]. Up to now, many new catalysts have been discovered, such as ZrO_2 [26], TiO_2 [24], Al_2O_3 [24], ZnO [27], nanodiamond [28, 29], C_{60} [29], and so on. In addition, Yao et al. reported the clone of CNTs by using an open CNT as the growth seed [30].

Distinct features were found for the CNTs grown from metal-free catalysts. For example, Liu et al. reported that the growth rate from SiO_x nanoparticles is 300 times lower than that of traditional Fe group metals. Therefore, CNTs as short as 150 nm could be obtained [31]. Homma et al. reported that the cap shapes of CNTs from nanodiamonds are polygonal rather than hemisphere [28]. These new phenomena strongly indicate that it is necessary to develop new growth mechanisms for the metal-free catalysts systems.

Currently, there are controversial results about the growth mechanisms for metal-free catalysts. Some critical questions such as the active species during the growth are still disputed. Liu et al. and Huang et al. used XPS to characterize the composition after CNTs growth and reported silicon oxide as the active catalysts [23, 24]. Steiner et al. investigated the growth processes of CNTs from ZrO_2 by using in situ XPS and confirmed that the catalysts keep oxide state during the growth [26]. However, Bachmatiuk et al. found that silicon oxides became SiC after the CNT growth and suggested that SiC was the active phase [32]. Homma et al. reported that the catalysts are solid during the CNTs growth because of their high melting points [21], while Huang et al. proposed that the particles are liquid because of a size-dependent melting point reduction [24].

From above results, we can see that as a new prosperous direction, there are still many unsolved problems for the CNTs growth with metal-free catalysts, such as the active phase, carbon atoms diffusion route, and so on. Therefore, in this chapter, the nucleation of CNTs with metal-free SiO_x as catalyst is investigated. The active composition, physical state of the catalyst particles, and possible carbon diffusion routes are compared with traditional Fe group catalysts so as to get a general understanding of the nucleation mechanism.

3.2 Experimental Design of MWCNT Nano-Furnaces for In Situ Observations

Currently, CVD has become the main method for controllable synthesis of CNTs. In CVD, the high temperature is provided by a resistance furnace; carbon source is provided by the decomposition of hydrocarbon gases; and catalysts are usually loaded on substrates. In this thesis, a new method is designed to mimic the CVD growth conditions and study the nucleation processes by using in situ TEM. A MWCNT fabricated by AAO template method acts as the carbon tubular furnace; high temperature is provided by Joule heating with a current passing the CNT; catalysts are loaded in the carbon tubular furnace; and defective carbon tubular walls are irradiated by electron beam to provide carbon source (Fig. 3.1).

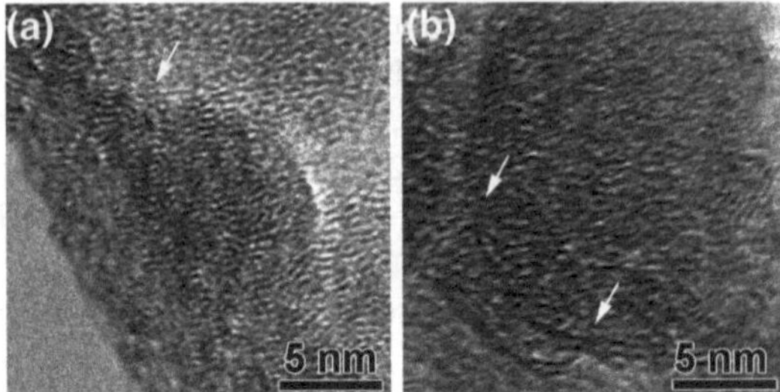

Fig. 3.1 In situ HRTEM observations on the surface diffusion (**a**) and lift-up (**b**) of carbon shells on SiO_x particles

The space in the pole piece of the TEM is limited to around 1 mm. Penetration depth of the electron beam is in the range of 1 μm. In addition, to acquire a HRTEM image, the thickness of the sample is restricted to less than ~100 nm. Therefore, a nanoscale growth furnace is needed. In this thesis, a MWCNT is used as the furnace. Its nanoscale diameter and light element are advantageous to the HRTEM observations. The carbon tubulars are produced by AAO template method as described in Chap. 2.

High temperature is necessary for the growth of CNTs to provide driving force for the carbon atoms diffusion, rearrangements, and also keep the catalyst active. In this thesis, we choose carbon tubulars fabricated by AAO template method with poor crystalline order and high electrical resistance. A current is passed through the CNT and high temperature is provided by the resulting Joule heating effect.

In CVD method, carbon sources are provided by hydrocarbon gases. And gas could be introduced into environmental TEMs to mimic the growth conditions [10, 12, 33–36]. However, gas is forbidden in common TEMs to keep the high vacuum (10^{-6} Pa). In this thesis, we mainly pay attention to the state of the catalyst and the interaction of the catalyst with carbon shells, rather than the decomposition process of the carbon precursors. And two methods are designed to provide carbon sources from solid phase precursors.

1. Put the catalyst in direct contact with the carbon tubular furnace. For catalyst such as Fe with strong interactions with carbon, the disordered carbon structures could act as the carbon sources. Carbon atoms surrounding the catalyst could be dissolved into the catalyst and precipitated as CNTs.
2. For catalysts having weak interactions with carbon, such as silicon oxide, electron beam irradiation is used to inject carbon atoms by inelastic scattering and provide carbon sources. By using this method, free carbon atoms or clusters could be generated and then absorbed on the surface of the catalyst and possibly rearranged into graphitic networks and nanotubes.

Catalyst particles with suitable size are critical for the nucleation and growth of CNTs. For example, it is believed that catalyst smaller than 5 nm are able to grow SWCNTs. In this thesis, size-tunable catalysts are filled into carbon tubular furnaces fabricated by AAO template method as described in Chap. 2. Fe and SiO_x nanoparticles are produced by thermal decomposition and CVD method, respectively.

3.3 Investigations of CNTs Nucleation Mechanism on SiO_x Catalysts Using In Situ TEM

SiO_x is a material previously being considered as inert and usually used as the substrate for CNTs growth. In 2009, it was surprisingly discovered that silicon oxide (SiO_x), as a metal-free material, could grow CNTs with high efficiency [23–25]. In this section, the nucleation behaviors of CNTs on SiO_x particles are investigated by using in situ TEM.

3.3.1 Surface Diffusion of Carbon and Lift-Up of Carbon Shells

First, amorphous SiO_x nanoparticles were uniformly filled into amorphous carbon tubular furnaces, as described in Chap. 2. When a current (10 μA) passed through the carbon tubular furnace and electron beam was irradiated on to the samples, the particles were found to be coated with carbon shells. The migration started from the MWCNT furnace wall as indicated by the HRTEM image (Fig. 3.1a). In the meantime, no obvious change was observed for the SiO_x nanoparticle, suggesting that the migration was only limited to the surface. The self-assembly of carbon atoms on surface suggests some interactions between carbon atoms and the SiO_x nanoparticles. As the carbon shells on the particle were accumulated, lift-up of a carbon cap was observed as marked by the arrows in Fig. 3.1b, which was considered as the key step for the nucleation of CNTs. The lift-up suggests that the interaction between the carbon shells and the SiO_x nanoparticles is not too strong so they can be separated. Therefore, our in situ observations confirm that carbon migration on surface, assembly of graphitic networks, and nucleation of CNTs on amorphous SiO_x nanoparticles are possible.

3.3.2 Nucleation of SWCNTs on Small SiO_X Nanoparticles

It was found that the structural evolutions of carbon shells on SiO_x nanoparticles are strongly dependent on the size of the nanoparticle. Small CNTs, even SWCNTs could be nucleated from small SiO_x nanoparticles. One example is presented in Fig. 3.2. Initially, an amorphous SiO_x nanoparticle of 4 nm in diameter was located close to the CNT furnace inner wall, which was also amorphous as revealed by the random and homogeneous contrast (a). Under high temperature and electron beam irradiation, the carbon shells became ordered indicating the self-assembly process. In the meantime, a single-walled carbon cap was observed on the protruded position of the nanoparticle (b), as dotted by the white spots. Another example is shown in Fig. 3.2c, where a SWCNT was grown attached to

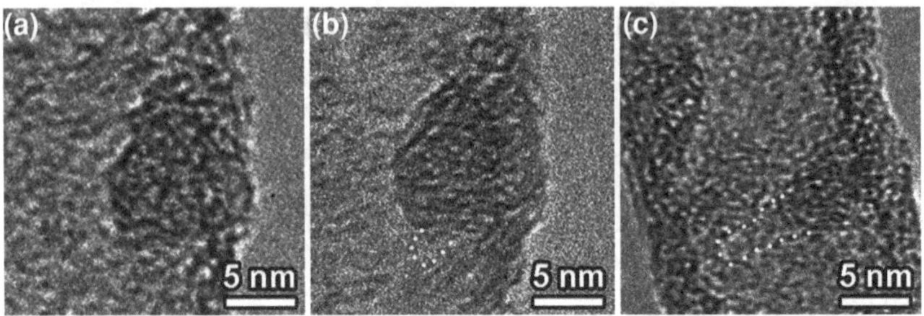

Fig. 3.2 TEM images of the nucleation and growth of CNTs on SiO_x particles: **a** Original SiO_x particle. **b** Formation of carbon cap. **c** Formation of a carbon nanotube. Reprinted with the permission from Ref. [37]. Copyright 2011, American Chemical Society

a small SiO_x nanoparticle. During the nucleation and growth processes, there was no change in the shape and size of the catalyst particle, suggesting that the catalyst particle kept solid. Therefore, it is revealed from our in situ observations that the active catalyst for metal-free catalyzed CNTs growth is solid and amorphous SiO_x.

3.3.3 *High Temperature Transformation of Large SiO_X Nanoparticles*

For large SiO_x nanoparticles (>20 nm), although carbon shells could be observed on the surfaces, no CNT could be nucleated. Under high temperature, reactions between SiO_x and carbon were observed. Initially, SiO_x nanoparticles were attached to the inner wall of a CNT furnace as shown in Fig. 3.3a–b. Under a current of 200 μA, SiO_x nanoparticles were found to be reduced to Si as revealed by the HRTEM image (Fig. 3.3c–d). When the current was further increased to 400 μA, Si nanoparticles were melted and finally reacted with surrounding carbon to SiC (Fig. 3.3e, f). During the heating and reaction processes, although the current and corresponding temperature were higher than previous experiments in Fig. 3.1,3.2, no carbon shells were found on the surfaces of Si or SiC nanoparticles, indicating that Si and SiC were not active materials for the assembly of carbon shells, that is, graphitization. These results further confirm that the active species for the metal-free CNTs growth from silicon oxide is at oxide state, rather than reduced or carbide states. Tsukimoto et al. investigated the reactions of SiO_2 by using in situ heating TEM stages [38], and found that SiO_2 was reduced to Si at the temperature of 1100–1200 °C. Wang et al. studied the solid state reactions between Si and C by using the same technique, and observed the formation of SiC at the temperature of 1400–1500 °C [39]. Usually, the CVD growth temperatures for CNTs are around or lower than 900 °C [23, 24], where SiO_2 is the stable phase. Therefore, our results are in consistence with these previous reports.

In addition, first principles calculations were conducted to investigate the interactions of CH_x clusters with Si and Si–O surfaces. It was found that all the CH_x clusters are chemically absorbed on the Si–O surface, while they can only absorb on

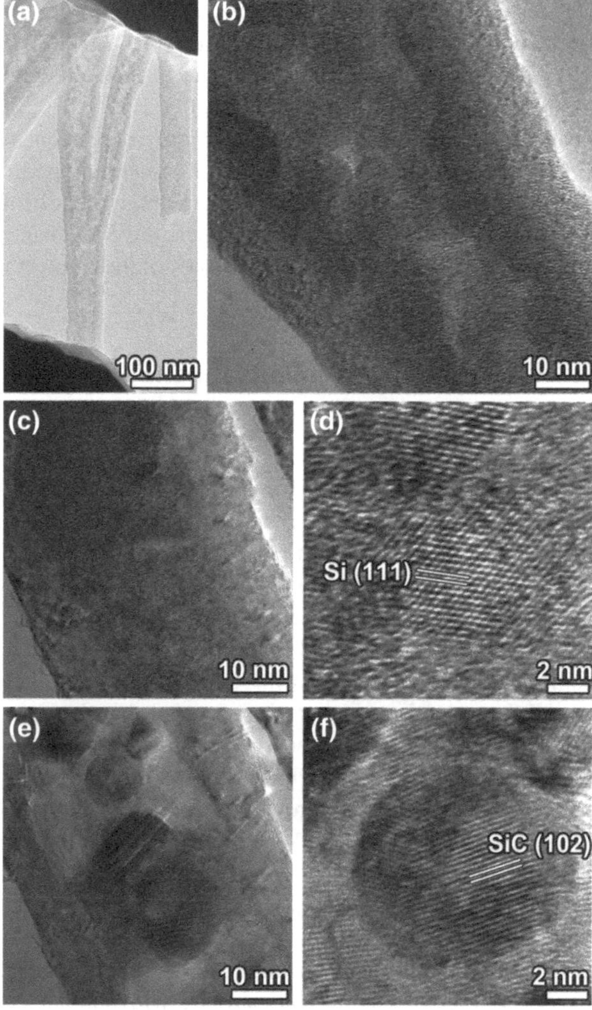

Fig. 3.3 In situ TEM observation on the structural evolution of the filled SiO_x particles under a high current. **a–b** Original SiO_x particles filled in a CNT. **c–d** Formation of Si nanocrystals. **e–f** Formation of SiC nanocrystals. Reprinted with the permission from Ref. [37]. Copyright 2011, American Chemical Society

the Si surface by physical adsorption with much lower adsorption energy. Especially, CH_4 is only weakly and physically absorbed on the Si surface, which is considered as the first important step for CVD. In addition, the comparison of growth behaviors of CNTs under identical conditions by Si and SiO_x nanoparticles confirms that while SiO_x have high efficiency, Si could not catalyze the CNTs growth. Therefore, both the theoretical calculations and CVD growth are consistent with the in situ TEM observations about the nature of the catalysts. Ding et al. studied the influence of interactions between carbon atoms and different metals by MD simulations, and found that carbon caps could only form on metals with strong enough interactions [40]. If the interaction is too weak, the carbon networks tend to form carbon ashes. The interactions between carbon and the catalysts are sensitive to the composition.

Our results demonstrate that for metal-free catalyst systems, the composition and interactions with carbon atoms are as important as those of the metal catalysts.

From above results, we could come to the following conclusions for the SiO_x catalyzed CNTs growth. First, the active catalyst is SiO_x, rather than SiC or Si. Second, the catalytic abilities are strongly dependent on the size of the catalyst. CNTs can only nucleate and grow on small nanoparticle, while SiO_x nanoparticles can assist the graphitization process regardless of the sizes. Third, during the CNTs nucleation and growth processes, the SiO_x nanoparticles are at solid state and carbon atoms diffuse on the surfaces of the catalyst nanoparticles.

3.4 General Requirements for the Catalysts Used for CNTs Growth

What are suitable catalysts for the growth of CNTs? In traditional VLS mechanism, the catalysts were described as a metal that can catalyze the decomposition of hydrocarbon gases and form liquid carbide phase. And the diameters of the CNTs are determined by the size of the catalyst particles. However, after the discoveries of new metal and metal-free catalysts, these questions should be reconsidered.

For the newly developed metal catalysts, Homma et al. proposed that the dissolution of carbon in metals increases at nanoscale [21], and the growth follows the VLS mechanism. Liu et al. suggested that the catalyst is only a template for the decomposition of hydrocarbon gases, assembly of the carbon networks, and nucleation of CNTs [20], which is based on the fact that many different kinds of metals could catalyze the growth of CNTs. For both mechanisms, we can see that the size of the catalysts is a key parameter, while the formation of carbide phase is not a required property.

For the metal-free catalysts, Liu et al. and Huang et al. observed nanoparticles around 2 nm on the substrates that are suitable for CNTs growth [23, 24]. Homma et al. characterized the semiconductor nanoparticles such as Ge and Si, and found that CNTs are attached to nanoparticles with similar sizes [22]. In addition, it was reported that CNTs grown on nano-diamonds are polygonal rather than hemisphere [28]. Therefore, the size of the catalysts or protrusion is also important for the metal-free catalysts systems.

In this thesis, it is found that the behaviors of carbon atoms are distinct on SiO_x nanoparticle with different sizes. For SiO_x nanoparticles larger than 20 nm, carbon networks self-assemble and encapsulate the whole particle. For the small SiO_x nanoparticles (~3 nm), the carbon caps are lifted up from the protruded area and grow into CNTs. Therefore, the diameter of the CNT is determined by the size of the protrusion, well consistent with previous assumptions and ex situ observations [28].

From above discussions, we can draw a conclusion that suitable size is commonly required for the catalysts used for CNTs growth. Another factor is the interaction between the catalyst and carbon. As previously mentioned, Ding et al. reported that strong interactions are essential for the nucleation of CNTs [40]. In

Table 3.1 Comparison of Fe and SiO_x catalysts used for CNTs growth

	Fe	SiO_x
Catalytic decomposition	Yes	No
Active catalyst	Fe_3C	SiO_x
Catalyst state	Liquid or quasi-liquid	Solid
Diffusion route	Bulk or surface layer	Surface
Growth mechanism	VLS	Solid surface growth

our experiments, it is found that with similar length scales, while carbon shells and CNTs could be grown on the surface of SiO_x nanoparticles, no graphitic structures could be assembled on the surfaces of Si and SiC nanoparticles. Therefore, we propose that suitable interaction between catalyst and carbon atoms is another common requirement for the catalyst of CNTs growths.

In Table 3.1, the characteristics of Fe- and SiO_x-based catalysts are compared, including the composition, nature, and state of active catalyst, ability to catalyze decomposition of hydrocarbon gases, and carbon atoms diffusion routes. The two catalysts are different in all the features mentioned above. Therefore, these are not the general requirements.

To summarize, we propose that the two common requirements for catalysts used for CNTs growth are suitable interactions with carbon and suitable size. Based on this understanding, we further investigate the growth of CNTs from all-carbon growth seeds, and heteroepitaxial growth from boron nitride nanofibers in the following sections.

3.5 CNTs Growth from All-Carbon Growth Seeds

In the last section, we propose two common requirements for catalysts for CNTs growth, i.e., suitable interactions with carbon and suitable particle size. Actually, the interaction between carbon atoms in graphitic networks is one of the strongest bonds in nature. And the CNTs in the first paper published by Dr. Iijima were synthesized by arc discharge method without using any heterocatalyst [41]. Therefore, in this section we use an all-carbon seed for the growth of CNTs [42], to study the structural evolutions of a carbon microfiber.

3.5.1 Structural Evolutions of Carbon Microfibers Under High Currents

A single carbon microfiber is connected between two Au electrodes of the in situ TEM-STM holder, as shown in Fig. 3.4a. It is kept at low current for about half an hour to get stable contact. After that, a bias up to 5 V is applied, and the current

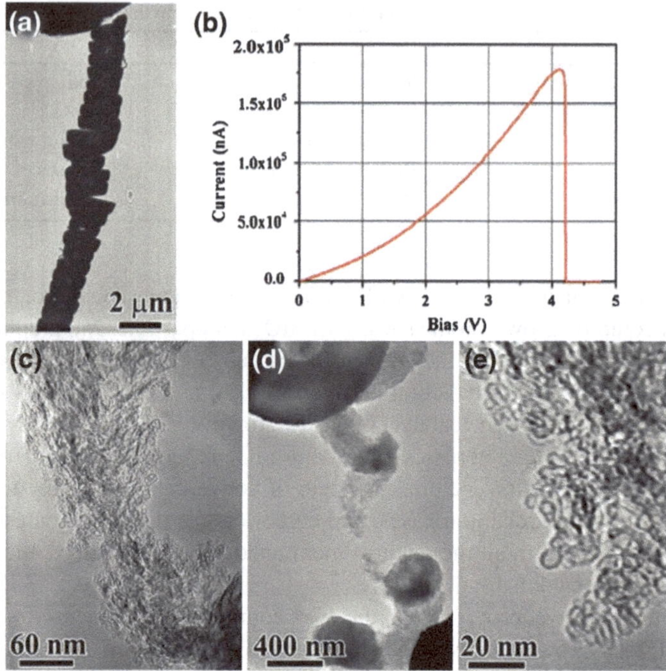

Fig. 3.4 Structural evolution of a carbon microfiber induced by a high-density direct current: **a** A carbon microfiber bridged between gold electrodes. **b** I–V curve recorded for the experimental system. **c–e** TEM images show that the resulting nanostructures after the carbon microfiber is broken. Reprinted from Ref. [42], Copyright 2009, with permission from Elsevier

increases approximately linearly, as shown in Fig. 3.4b. The current reaches about 175 μA and suddenly drops to zero, when the bias is about 4.1 V. TEM images show that the carbon microfiber is damaged (Fig. 3.4c–e) along with obvious shrinkage of the diameter and formation of hollow nanostructures.

The products are transferred to TEM grids and further characterized by high resolution TEM. As shown in Fig. 3.5, hollow graphitic structures with diameters from 4–10 nm are revealed (a, b). Detailed characterizations show that small structures such as single-walled and double-walled fullerenes (c, d). More importantly, double-walled CNTs with diameters around 4 nm and length about 10 nm are observed (e, f). No catalysts could be found at the root or the end of the CNT. And the walls of the CNT are connected to the graphitic layers of the carbon microfiber.

3.5.2 Formation Mechanism of CNTs from Carbon Microfibers

Because there are no catalysts found attached to the grown CNT, the formation mechanism should be different from traditional catalytic growth. Previously, it was

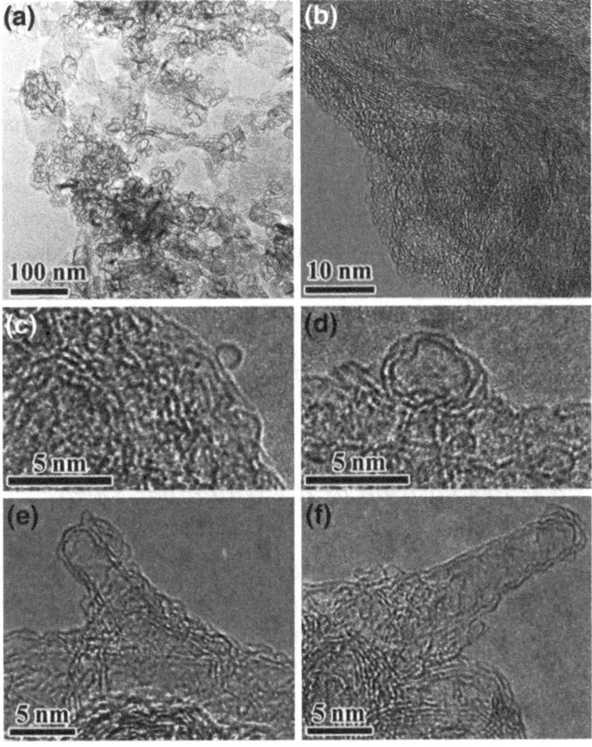

Fig. 3.5 Images of the resulting product from the carbon microcoil: **a** Low-magnification image. **b** HRTEM image of hollow graphitic spheres. **c**– **d** HRTEM images of fullerenes. **e**–**f** HRTEM images of double-walled carbon nanotubes. Reprinted from Ref. [42], Copyright 2009, with permission from Elsevier

reported that after high temperature, the carbon microfibers became ordered and denser rather than porous [43, 44]. Therefore, the high temperature induced by Joule heating is not the only driving force. It is noteworthy that the current density is as high as 10^4 A/cm^2, which might lead to electromigration effects [45–47]. Indeed, obvious mass loss is observed after the breaking of the microfiber at high current. Consequently, we propose that the formation of hollow graphitic structures, including graphitic hollow spheres, fullerenes, and CNTs are the synergistic effects of both high temperature and high-density current. Carbon microfibers are composed purely of carbon. In addition, no impurities could be detected during and after the structural transition. Therefore, our in situ TEM observations confirm that with suitable driving force, CNTs could be self-assembled without the assistance of heterocatalysts, and the growth of all-carbon CNTs is possible.

As early as 1991, Iijima reported the discovery of CNTs produced by arc discharge method [41]. The smallest CNT was double walled with the diameter around 5.5 nm. Recently, in 2009, Takagi et al. and Rao et al. reported the synthesis of CNTs by using carbon growth seeds such as nanodiamonds and C_{60}, respectively [28, 29]. Also in 2009, Yao et al. cloned CNTs by using open-CNTs as growth seeds [30]. These reports are consistent with our in situ TEM results, and confirm our assumptions about the required conditions for the catalysts used for CNTs growth.

3.6 Heteroepitaxial Growth of CNTs

Hexagonal boron nitride (h-BN) has a similar structure with graphite. In addition, previous works show that boron and nitrogen atoms have strong interactions with carbon. These three elements can form compound nanostructures such as zero-dimensional fullerenes, one-dimensional nanotubes, and two-dimensional graphene structures [48–51]. According to the proposed required conditions for CNTs growth, if h-BN nanostructures with suitable size could be produced, they have the potential to be used as growth seeds. What is more, the structure and chemical properties of h-BN are very stable; therefore, h-BN may be advantageous to the structure control of the grown CNTs. In this section, BN nanofibers are produced and used as growth seeds for CNTs. It is found that CNTs grow following an epitaxial growth mechanism.

3.6.1 Preparation of BN Nanofiber Seeds

A floating catalyst CVD method is used for the synthesis of BN nanofibers [52, 53]. The following are the experimental conditions: growth temperature is 1350 °C, ferrocene is the precursor for catalysts, NH_3 is used as nitrogen source, boron and B_2O_3 mixture is used as boron source, mixture of Fe_2O_3 and FeS is used as growth

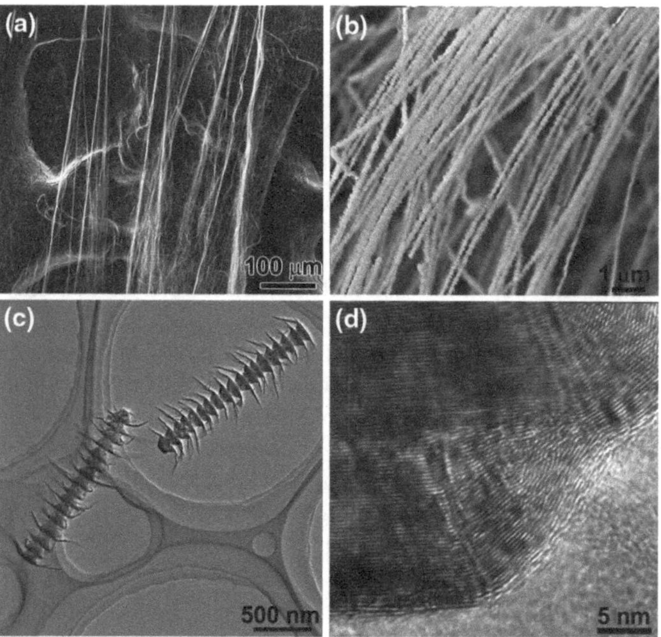

Fig. 3.6 Structural characterizations of the boron nitride nanofibers: **a**, **b** SEM images **c**, **d** TEM and HRTEM images

promoter, Ar is used as carrier gas, and the growth time is 30–60 min. After the growth, white products are collected in the middle of the reaction zone.

Structure characterizations are shown in Fig. 3.6. SEM images reveal the high purity of the products (a, b). The BN nanofibers are as long as millimeter and well aligned. The diameters of the nanofibers are about 200 nm. On the surfaces, many sharp protrusions could be seen. TEM images show that the nanofibers are composed of cup-stacked sections, of which the thickness is around 100 nm (c, d). HRTEM image reveals that the core part of the nanofibers are composed of BN (002) planes, stacking perpendicular to the growth direction. At the edges, BN (002) planes are extended to form nanometer scale tips.

3.6.2 Growth of SWCNTs from BN Nanofiber Seeds

First, the produced BN nanofibers are soaked in concentrated hydrochloric acid and centrifuged to remove remaining metallic contaminations. After that, the BN nanofibers are loaded on Si substrates. Then SWCNTs are grown following a typical approach for metal-free catalyst [23]. The growth temperature is 900 °C, CH_4 is used as carbon source, and the growth lasts 10–15 min. The products are characterized by Raman spectroscopy, SEM, and TEM.

SEM images show that abundant nanotubes are grown along and across the BN nanofibers (Fig. 3.7a–c). Higher magnification SEM image demonstrates that the nanotubes are rooted on the surface of the BN nanofibers and connected with the edges of the platelets (Fig. 3.7d). Different from nanotubes grown from metallic catalysts, the nanotubes grown from BN nanofibers are mostly isolated or in small bundles.

Figure 3.8 shows a Raman spectrum of the product. G mode of BN at 1366 cm^{-1}, G and D mode of CNTs at 1590 cm^{-1} and 1322 cm^{-1} could be clearly identified. In addition, RBM band of SWCNTs at 181 cm^{-1} and 129 cm^{-1} are detected. The RBM band could be used to calculate the diameters of the SWCNTs by using the following formula [55–57]: $\omega_{RBM} = 223.5/d_t + 12.5$, where ω_{RBM} is the Raman shift of the RBM peak and d_t is the diameter of the nanotube. The diameters of the SWCNTs are calculated to be mainly at 1.32 nm and 1.92 nm. It is interesting that the diameters of the nanotubes are about four and six times of graphite and h-BN (002) interplanar distances. In last section, it is revealed that the BN nanofibers are composed of platelets with (002) planes perpendicular to the fiber axis. It is possible that there is certain dependence of the nanotube diameters on the edge structures of the BN nanofibers.

The products are transferred onto TEM grids for further characterizations by using HRTEM, as shown in Fig. 3.9. A small bundle of two nanotubes are grown along the surface of a BN nanofiber (a), with a diameter of about 2.0 nm. Another small bundle of two nanotubes is shown in (b), which clearly reveals that the nanotubes are rooted on the BN fiber and the walls are connected with the (002) planes of the protruding edges. The end of a SWCNT is presented in (c), as indicated by the white circle, where no catalyst particle could be seen. Above observations confirm that the BN nanofibers are the growth seeds for the nanotubes.

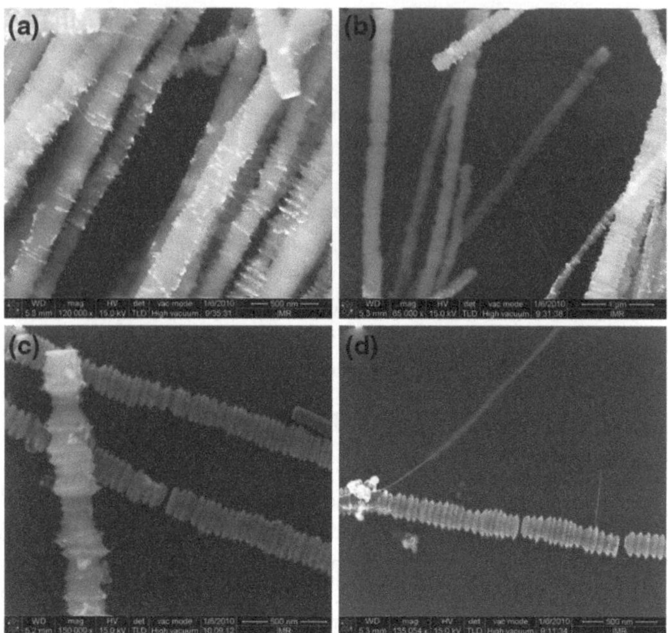

Fig. 3.7 SEM images of the SWCNTs synthesized by BN nanofibers [54]

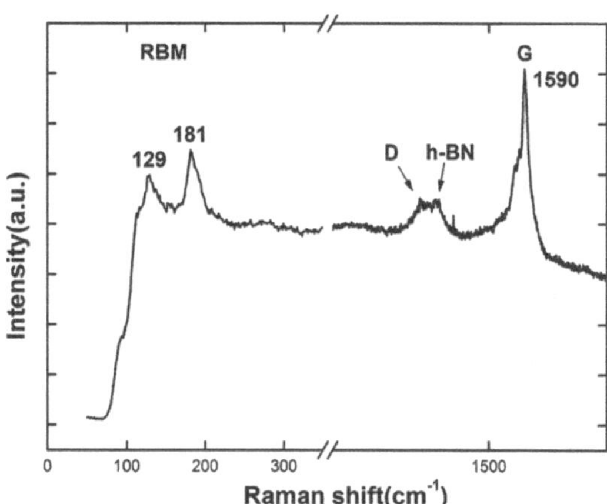

Fig. 3.8 Raman spectrum of the CNTs grown from BN nanofibers [54]

3.6.3 Growth Mechanism of CNTs from BN Nanofibers

h-BN is highly stable and can resist oxidation when the temperature is up to 900 °C in air [58]. Therefore, the catalysts during the CNTs growth are BN rather than oxide or carbide, which is confirmed by HRTEM images (Fig. 3.9). In addition, both Raman spectroscopy and HRTEM images reveal that the diameters of

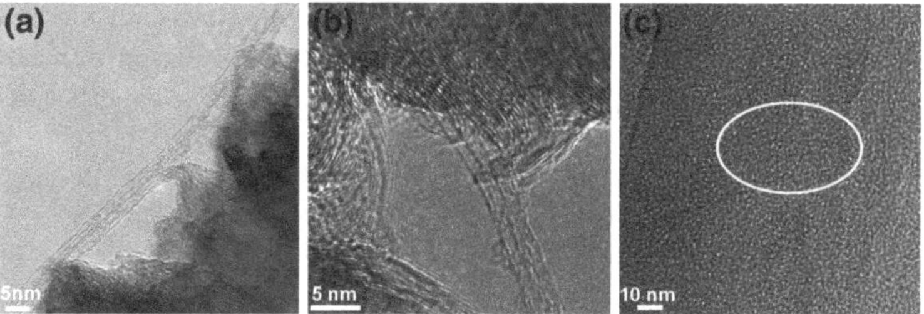

Fig. 3.9 TEM images of the SWCNTs synthesized from boron nitride nanofibers

the SWCNTs are integer multiples of BN (002) interplanar distances. And no other catalyst particles could be detected at the roots or tips of the CNTs. We propose that SWCNTs grown from BN nanofibers seeds following a heteroepitaxial mechanism, and therefore they have fixed orientation and lattice matching relationships. In semiconductor industry, epitaxy is a mature technique and has widespread applications [59]. By using this technique, it is possible to control the crystalline structure, orientation, and defects concentration because of the orientation and lattice matching relations. Recently, the development of heteroepitaxial growth makes large-scale production of graphene possible [60]. The heteroepitaxial growth developed in this thesis provides a new approach for the precise control of CNTs diameters and may offer implications to realize the chirality-controlled growth.

3.7 Summary

In this chapter, the growth mechanisms of CNTs are investigated by using in situ TEM method. A general requirement for the catalyst is proposed, based on which CNTs growth from all-carbon seeds and epitaxial growth from BN nanofibers are explored. The following are the main conclusions:

1. The catalysts used for metal-free SiO_x catalyzed CNTs growth are at amorphous, solid, and oxide state during the growth. And carbon atoms are provided by surface diffusion. By comparing with metallic Fe-based catalysts, a common condition required for the catalysts used for CNTs growth is proposed, that is, suitable particle size and suitable interactions with carbon. The ability to catalytically decompose hydrocarbon gases and the formation of liquid carbide phase are not indispensable.
2. The growth of CNTs from all-carbon seeds are investigated by passing a high-density current through carbon microfibers and the structural evolutions are observed in realtime. The carbon microfibers are transformed into various hollow structures such as graphitic spheres, fullerenes, and CNTs. The in situ observations confirm that it is possible to grow CNTs without using heterocatalysts.
3. BN nanofibers are synthesized and used as growth seeds for CNTs. Due to the structural similarity and strong interactions between boron, nitrogen, and carbon

atoms. CNTs are grown following a heteroepitaxial mechanism. As a result, the diameters of the SWCNTs are multiples of the interplanar distances of BN (002) planes, providing a new approach for the structure controllable synthesis of CNTs.

References

1. Wagner RS, Ellis WC (1964) Vapor-liquid-solid mechanism of single crystal growth. Appl Phys Lett 4:89
2. Baker RTK, Barber MA, Harris PS, Feates FS, Waite RJ (1972) Nucleation and growth of carbon deposits from the nickel catalyzed decomposition of acetylene. J Catal 26:51–62
3. Baker RTK, Gadbsy GR, Terry S (1975) Formation of carbon filaments from catalysed decompasition of hydrocarbons. Carbon 13:245–246
4. Tibbetts GG (1984) Why are carbon filaments tubular? J Cryst Growth 66:632–638
5. Dai H, Rinzler AG, Nikolaev P, Thess A, Colbert DT, Smalley RE (1996) Single-wall nanotubes produced by metal-catalyzed disproportionation of carbon monoxide. Chem Phys Lett 260:471–475
6. Li F (2001) Synthesis and physical properties of single-walled carbon nanotubes by catalytic decomposition of hydrocarbons. PhD thesis. Chinese Academy of Science, Shenyang
7. Ren W-C (2005) Controllable synthesis, growth mechanism and physical properties of carbon nanotubes. PhD thesis. Academy of Sciences, Shenyang
8. Huang SM, Woodson M, Smalley R, Liu J (2004) Growth mechanism of oriented long single walled carbon nanotubes using "fast-heating" chemical vapor deposition process. Nano Lett 4:1025–1028
9. Huang JY (2007) In Situ observation of quasimelting of diamond and reversible graphite — diamond phase transformations. Nano Lett 7:2335–2340
10. Helveg S, López-Cartes C, Sehested J, Hansen PL, Clausen BS, Rostrup-Nielsen JR, Abild-Pedersen F, Nørskov JK (2004) Atomic-scale imaging of carbon nanofibre growth. Nature 427:426–429
11. Hofmann S, Blume R, Wirth CT, Cantoro M, Sharma R, Ducati C, Hävecker M, Zafeiratos S, Schnoerch P, Oestereich A, Teschner D, Albrecht M, Knop-Gericke A, Schlögl R, Robertson J (2009) State of transition metal catalysts during carbon nanotube growth. J Phys Chem C 113:1648–1656
12. Yoshida H, Takeda S, Uchiyama T, Kohno H, Homma Y (2008) Atomic-scale in situ observation of carbon nanotube growth from solid state iron carbide nanoparticles. Nano Lett 8:2082–2086
13. Takagi D, Homma Y, Hibino H, Suzuki S, Kobayashi Y (2006) Single-walled carbon nanotube growth from highly activated metal nanoparticles. Nano Lett 6:2642–2645
14. Bhaviripudi S, Mile E, Steiner SA, Zare AT, Dresselhaus MS, Belcher AM, Kong J (2007) CVD synthesis of single-walled carbon nanotubes from gold nanoparticle catalysts. J Am Chem Soc 129:1516–1517
15. Takagi D, Kobayashi Y, Hlbirio H, Suzuki S, Homma Y (2008) Mechanism of gold-catalyzed carbon material growth. Nano Lett 8:832–835
16. Zhou W, Han Z, Wang J, Zhang Y, Jin Z, Sun X, Zhang Y, Yan C, Li Y (2006) Copper catalyzing growth of single-walled carbon nanotubes on substrates. Nano Lett 6:2987–2990
17. Ritschel M, Leonhardt A, Elefant D, Oswald S, Büchner B (2007) Rhenium-catalyzed growth carbon nanotubes. J Phys Chem C 111:8414–8417
18. Liu B, Ren W, Gao L, Li S, Liu Q, Jiang C, Cheng H-M (2008) Manganese-catalyzed surface growth of single-walled carbon nanotubes with high efficiency. J Phys Chem C 112:19231–19235
19. Zhang Y, Zhou W, Jin Z, Ding L, Zhang Z, Liang X, Li Y (2008) Direct growth of single-walled carbon nanotubes without metallic residues by using lead as a catalyst. Chem Mater 20:7521–7525

20. Yuan D, Ding L, Chu H, Feng Y, McNicholas TP, Liu J (2008) Horizontally aligned single-walled carbon nanotube on quartz from a large variety of metal catalysts. Nano Lett 8:2576–2579
21. Homma Y, Liu HP, Takagi D, Kobayashi Y (2009) Single-walled carbon nanotube growth with non-Iron-group "Catalysts" by chemical vapor deposition. Nano Res 2:793–799
22. Takagi D, Hibino H, Suzuki S, Kobayashi Y, Homma Y (2007) Carbon nanotube growth from semiconductor nanoparticles. Nano Lett 7:2272–2275
23. Liu B, Ren W, Gao L, Li S, Pei S, Liu C, Jiang C, Cheng H-M (2009) Metal-catalyst-free growth of single-walled carbon nanotubes. J Am Chem Soc 131:2082–2083
24. Huang S, Cai Q, Chen J, Qian Y, Zhang L (2009) Metal-catalyst-free growth of single-walled carbon nanotubes on substrates. J Am Chem Soc 131:2094–2095
25. Hirsch A (2009) Growth of single-walled carbon nanotubes without a metal catalyst-A surprising discovery. Angew Chem Int Ed 48:5403–5404
26. Steiner SA, Baumann TF, Bayer BC, Blume R, Worsley MA, MoberlyChan WJ, Shaw EL, Schlögl R, Hart AJ, Hofmann S, Wardle BL (2009) Nanoscale zirconia as a nonmetallic catalyst for graphitization of carbon and growth of single- and multiwall carbon nanotubes. J Am Chem Soc 131:12144–12154
27. Gao FL, Zhang LJ, Huang SM (2010) Zinc oxide catalyzed growth of single-walled carbon nanotubes. Appl Surf Sci 256:2323–2326
28. Takagi D, Kobayashi Y, Homma Y (2009) Carbon nanotube growth from diamond. J Am Chem Soc 131:6922–6923
29. Rao F, Li T, Wang Y (2009) Growth of "all-carbon" single-walled carbon nanotubes from diamonds and fullerenes. Carbon 47:3580–3584
30. Yao Y, Feng C, Zhang J, Liu Z (2009) Cloning of single-walled carbon nanotubes via open-end growth mechanism. Nano Lett 9:1673–1677
31. Liu B, Ren W, Liu C, Sun C-H, Gao L, Li S, Jiang C, Cheng H-M (2009) Growth velocity and direct length-sorted growth of short single-walled carbon nanotubes by a metal-catalyst-free chemical vapor deposition process. ACS Nano 3:3421–3430
32. Bachmatiuk A, Börrnert F, Grobosch M, Schäffel F, Wolff U, Scott A, Zaka M, Warner JH, Klingeler Rd, Knupfer M, Büchner B, Rümmeli MH (2009) Investigating the graphitization mechanism of SiO2 nanoparticles in chemical vapor deposition. ACS Nano 3:4098–4104
33. Sharma R, Iqbal Z (2004) In situ observations of carbon nanotube formation using environmental transmission electron microscopy. Appl Phys Lett 84:990–992
34. Sharma R, Rez P, Treacy MMJ, Stuart SJ (2005) In situ observation of the growth mechanisms of carbon nanotubes under diverse reaction conditions. J Electron Microsc 54:231–237
35. Lin M, Tan JPY, Boothroyd C, Loh KP, Tok ES, Foo YL (2006) Direct observation of single-walled carbon nanotube growth at the atomistic scale. Nano Lett 6:449–452
36. Hofmann S, Sharma R, Ducati C, Du G, Mattevi C, Cepek C, Cantoro M, Pisana S, Parvez A, Cervantes-Sodi F, Ferrari AC, Dunin-Borkowski R, Lizzit S, Petaccia L, Goldoni A, Robertson J (2007) In situ observations of catalyst dynamics during surface-bound carbon nanotube nucleation. Nano Lett 7:602–608
37. Liu B, Tang D-M, Sun C, Liu C, Ren W, Li F, Yu W-J, Yin L-C, Zhang L, Jiang C, Cheng H-M (2011) Importance of oxygen in the Metal-free catalytic growth of single-walled carbon nanotubes from SiOx by a vapor–solid–solid mechanism. J Am Chem Soc 133:197–199
38. Tsukimoto S, Sasaki K, Hirayama T, Saka H (1997) Vibration of an interface between Si and SiO2 during reduction of SiO2. Philos Mag Lett 76:173–179
39. Wang Y, Huang J, Jiao T, Zhu Y, Hamza A (2007) Abnormal strain hardening in nanostructured titanium at high strain rates and large strains. J Mater Sci 42:1751–1756
40. Ding F, Larsson P, Larsson JA, Ahuja R, Duan H, Rosen A, Bolton K (2007) The importance of strong carbon –metal adhesion for catalytic nucleation of single-walled carbon nanotubes. Nano Lett 8:463–468
41. Iijima S (1991) Helical microtubules of graphitic carbon. Nature 354:56–58
42. Tang DM, Liu C, Li F, Ren WC, Du JH, Ma XL, Cheng HM (2009) Structural evolution of carbon microcoils induced by a direct current. Carbon 47:670–674
43. Wu FY, Du JH, Liu CG, Li LX, Cheng HM (2004) The microstructure and energy storage characteristics of micro-coiled carbon fibers. New Carbon Mater 19:81–86

44. Wu FY, Cheng HM (2005) Structure and thermal expansion of multi-walled carbon nanotubes before and after high temperature treatment. J Phys D-Appl Phys 38:4302–4307
45. Svensson K, Olin H, Olsson E (2004) Nanopipettes for metal transport. Phys Rev Lett 93:145901
46. Regan BC, Aloni S, Ritchie RO, Dahmen U, Zettl A (2004) Carbon nanotubes as nanoscale mass conveyors. Nature 428:924–927
47. Heinze S, Wang N-P, Tersoff J (2005) Electromigration forces on ions in carbon nanotubes. Phys Rev Lett 95: 186802
48. Guo T, Jin CM, Smalley RE (1991) Doping bucky—formation and properties of boron-doped buckminsterfullerene. J Phys Chem 95:4948–4950
49. Miyamoto Y, Rubio A, Cohen ML, Louie SG (1994) Chiral tubules of hexagonal BC2 N. Phys Rev B 50:4976
50. Krivanek OL, Chisholm MF, Nicolosi V, Pennycook TJ, Corbin GJ, Dellby N, Murfitt MF, Own CS, Szilagyi ZS, Oxley MP, Pantelides ST, Pennycook SJ (2010) Atom-by-atom structural and chemical analysis by annular dark-field electron microscopy. Nature 464:571–574
51. Ci L, Song L, Jin C, Jariwala D, Wu D, Li Y, Srivastava A, Wang ZF, Storr K, Balicas L, Liu F, Ajayan PM (2010) Atomic layers of hybridized boron nitride and graphene domains. Nat Mater 9:430–435
52. Tang DM, Liu C, Cheng HM (2006) Platelet boron nitride nanowires. NANO 1:65–71
53. Tang DM, Liu C, Cheng HM (2007) Controlled synthesis of quasi-one-dimensional boron nitride nanostructures. J Mater Res 22:2809–2816
54. Tang DM, Zhang LL, Liu C, Yin LC, Hou PX, Jiang H, Zhu Z, Li F, Liu BL, Kauppinen EI, Cheng HM (2012) Heteroepitaxial growth of single-walled carbon nanotubes from boron nitride. Sci Report 2:971
55. Dames C, Chen S, Harris CT, Huang JY, Ren ZF, Dresselhaus MS, Chen G (2007) A hot-wire probe for thermal measurements of nanowires and nanotubes inside a transmission electron microscope. Rev Sci Instrum 78:104903
56. Bachilo SM, Strano MS, Kittrell C, Hauge RH, Smalley RE, Weisman RB (2002) Structure-assigned optical spectra of single-walled carbon nanotubes. Science 298:2361–2366
57. Huang JY, Ding F, Jiao K, Yakobson BI (2007) Real Time Microscopy, Kinetics, and Mechanism of Giant Fullerene Evaporation. Phys Rev Lett 99:175503
58. Chen Y, Zou J, Campbell SJ, Le Caer G (2004) Boron nitride nanotubes: pronounced resistance to oxidation. Appl Phys Lett 84:2430–2432
59. Chen S, Huang JY, Wang Z, Kempa K, Chen G, Ren ZF (2005) High-bias-induced structure and the corresponding electronic property changes in carbon nanotubes. Appl Phys Lett 87:263107–2631030
60. Huang JY, Chen S, Ren ZF, Wang ZQ, Wang DZ, Vaziri M, Suo Z, Chen G, Dresselhaus MS (2006) Kink formation and motion in carbon nanotubes at high temperatures. Phys Rev Lett 97:075501

Chapter 4
Fabrication and Property Investigation of Carbon Nanotube-Clamped Metal Atomic Chains

Metal atomic chains (MACs) are extremely small one-dimensional structures [1]. They have unique physical properties due to the special structures, such as quantized conductance, quantum magnetoresistivity, and so on [2–5]. Therefore, MACs based devices have been proposed, such as the quantum electronic switches and integrated circuits of quantum electronic logic units [6]. However, due to the ultra-small dimensions of MACs, the manipulation, connection, and fabrication are extremely challenging. As we know, CNTs are one-dimensional structures with excellent electrical properties [7]. Recently, the fabrication of CNTs based nanodevices is becoming mature, and applications such as ballistic quantum wires, field-effect transistors and integrated circuits have been reported [8–10]. In addition, it was found that CNTs could form strong covalent bonds with metals, and the interface between them has small scattering for electron and spin transport [11, 12]. In this thesis, a new hybrid structure is designed, i.e. CNT-clamped MACs, in which CNTs are used as nanoscale conducting wires for the connection of MACs.

4.1 Research Background

4.1.1 Fabrication Methods for MACs

There are three methods for the fabrication of MACs, i.e., STM, in situ TEM and mechanically controllable break junction (MCBJ) [13–15]. Essentially, these methods are based on detachment of metal contacts, either by mechanical force or electron beam irradiation. At the moment of breaking, a nanoscale even atomic bridge-like structure could be formed. The common point of the three methods is that bulk metals are used as the electrodes. Therefore, the device is of macro-scale size, which restricts their applications in nanoscale devices. Ohnishi et al. combined the in situ TEM and STM methods by introducing STM probes into TEM [3], therefore obtained high-resolution in situ observations during the fabrication

D.-M. Tang, *In Situ Transmission Electron Microscopy Studies of Carbon Nanotube Nucleation Mechanism and Carbon Nanotube-Clamped Metal Atomic Chains*, Recognizing Outstanding Ph.D. Research, DOI: 10.1007/978-3-642-37259-9_4,

of MACs and correlated the observations with the electrical properties. Quantized conductance of Au atomic chains has been demonstrated by this in situ method. However, the assembly of MACs at nanometer scale is still an unsolved problem.

4.1.2 Structures and Properties of MACs

Because of the atomic size, high surface-volume ratio and quantum effects, the MACs have unique structures and physical properties. For example, it was found that the interatomic distances of Au MACs are larger than those of bulk Au. Interestingly, peculiar tubular structures could be formed for the Au MACs. Even more magically, the atom numbers of the Au MACs form an arithmetic sequence, with 7 more atoms for each consecutive shell [16]. Because of the high ratio of surface atoms, the surface tension plays an important role in the phase transitions of the MACs. When the thickness is smaller than 8 atomic layers, the Au MAC is reconstructed and transformed from (001) to (111) orientation, of which the surface energy is lowest [14, 17].

Because of the extremely small size, the MACs are intrinsically defect-free. Therefore, the strengths of the MACs are up to 5–13 GPa, approaching the theoretical limits and much higher than those of their bulk counterparts [18, 19]. The mechanical properties were investigated with AFM combined with MD simulations by Rubio-Bollinger et al. [19]. Series of plastic deformations due to the reconstructions were recorded and the fracture force was measured to be ~1.5 nN, with the fracture strength 10 times higher than that of bulk Au. The electrical properties of the Au atomic chains were monitored along with the formation processes. Along with the reduction of the atomic layers, the electrical conductivity is decreased in a stepwise manner. And the steps are located at multiples of conductance quantum ($G_0 = 2e^2/h \sim 13\ k\Omega^{-1}$), which means that the conductivity of the MACs is quantized [3]. The quantized conductance can be described by the Landau equation [20], $G = G_0 \Sigma T_i$, where T_i is the tunneling probability corresponding to a conductance channel. In addition, due to lack of electron degeneracy, the electrons at Fermi levels of ferromagnetic metals are spin-polarized, and quantized magnetoresistance properties have been demonstrated [21].

4.1.3 Nanodevices Based on MACs

Because of the unique physical properties, nanodevices based on MACs have been proposed, such as quantized atomic switches, by using a gate to control the ON and OFF states [6]. In addition, multi-level atomic switches with superior stabilities have been developed to realize logic operations of multiple states. What's more, spin-filters making use of the half metallic ferromagnetic MACs could be used in nano-spintronic devices [5].

4.2 Design and Fabrication of CNT-Clamped MACs

Despite the abundant researches on the MACs, from mechanical, electronic and magnetic properties to potential applications in nanodevices, there lacks an efficient method for the connection and assembly of these extremely small structures. Taking advantages of the superior electrical properties of CNTs and the mature devices fabrication techniques, a CNT-clamped MAC hybrid structure is designed in this thesis to solve the problem. Metal filled CNTs are used as the starting materials. Electron beam irradiation is combined with the in situ manipulations of TEM-STM holder for the fabrication. And the fabrication is carried out inside the TEM with high level vacuum (~10^{-6} Pa) to protect the active metal surfaces. The formation mechanism is investigated by using HRTEM and first principles calculations. And the electrical properties are measured in real time to be correlated with the atomic structures.

The fabrication of CNT-clamped MACs includes the production of metal filled CNTs, selective removal of carbon shells, thinning of the exposed metal nanowires, and finally tensile elongation. The experimental procedure is demonstrated in Fig. 4.1.

4.2.1 Formation Processes of Fe Atomic Chains

As demonstrated in Fig. 4.2, initially a single-crystalline Fe nanowire of about 100 nm in length was filled in the CNT, with the outer and inner diameters of ~25

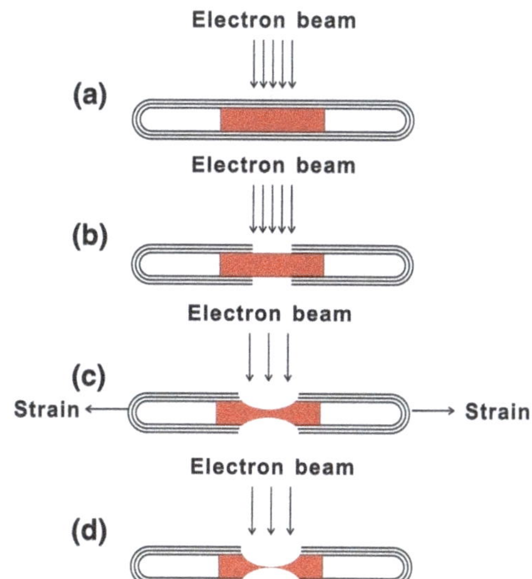

Fig. 4.1 Schematic of the fabrication process of CNT-clamped MACs: **a** Production of metal filled CNTs as the starting material, **b** peeling off carbon shells by electron beam irradiation induced knock-on effect, **c** thinning of metal nanowire by electron beam irradiation associated surface sputtering effect, **d** elongation and formation of MACs by tensile strain

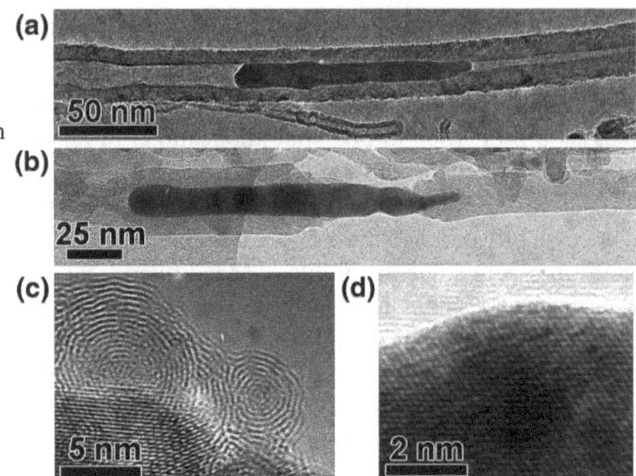

Fig. 4.2 The formation of a CNT-clamped Fe nanowire. **a** The original Fe filled CNT, **b** the CNT-clamped Fe nanowire after strong electron beam irradiation, **c** the reconstructed carbon onions from the peeled carbon shells, and **d** HRTEM image of the exposed Fe nanowire surface [22]

and ~10 nm, respectively. When the electron beam (100–300 A/cm^2) was focused on the filling position, and scanned back and forth, carbon shells were removed due to the knock-on effects of the electron irradiation. HRTEM images of the exposed Fe nanowire revealed that it remained single-crystalline, and carbon shells were removed and transformed into carbon onions at the boundaries of the electron beam, confirming the elemental selectivity of electron irradiation. After that, electron beam was focused on the surface of the Fe nanowire. Due to a surface sputtering effect, the exposed Fe nanowire was thinned with a much slower speed.

The structural transformation during the thinning process of a Fe nanowire under electron beam was investigated in detail, as shown in Fig. 4.3. After strong electron beam irradiation, carbon shells have been removed and the Fe crystal has been thinned as revealed by the brighter contrast in the central area (Fig. 4.3a). A hexagonal shaped hole appeared in the center, and was surrounded by a thin area. HRTEM image shows that the surface of the hole was composed of (110) planes as revealed by the lattices with interplanar distance of ~0.2 nm. The surface is distinct from face-centered cubic (FCC) metals, which are mainly composed of (111) planes [14, 17].

When the thickness of the Fe crystal was reduced to a certain threshold value, reorientation was observed as shown in Fig. 4.4. Initially, the Fe crystal was single crystalline with [111] zone axis as indicated by the hexagonal stacking order in surrounding areas. The central part was transformed into a square-shaped lattices, corresponding to the [100] zone axis (Fig. 4.4a). The reconstruction could be explained by the surface tension induced reorientation as simply explained by the two dimensional model (Fig. 4.4b). The orientation of the initial and transformed areas could be described as $[1\text{-}10]_{(111)}//[1\text{-}10]_{(001)}$, $[11\text{-}2]_{(111)}//[110]_{(001)}$.

When the diameter of the Fe nanowire was reduced to ~6 nm, the density of the electron beam was decreased to normal level for HRTEM observations (10–30 A/cm^2), so as to slow down the etching rate and record the structural evolutions. Because of the thermal gradient and corresponding thermal stress, the small Fe nanowire was continuously stretched and shrunk until finally the formation of an

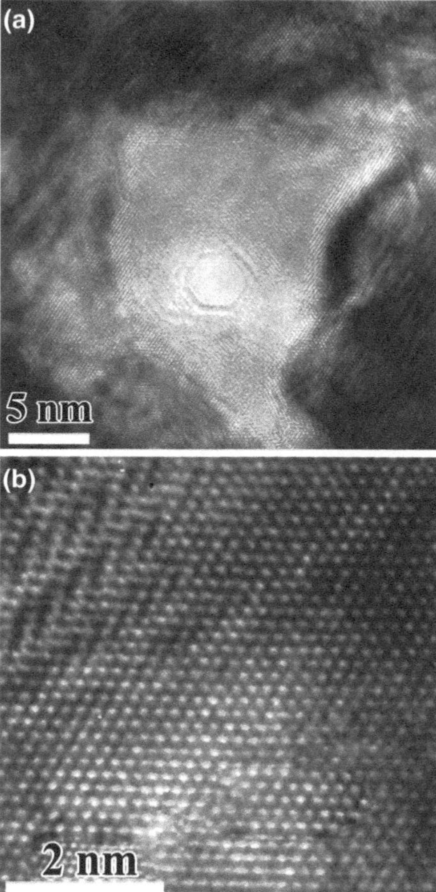

Fig. 4.3 HRTEM characterizations of the thinning process of a Fe single crystal by using strong electron irradiation. **a** TEM image of the formed hexagonal hole in the central irradiated area. **b** HRTEM image of the thin area with lattices corresponding to Fe (110) plane

atomic chain, as presented in Fig. 4.5. During the formation process, atomic steps along (110) planes were observed (c, d), and distortion-recovery was found (e, f), indicating slip is an important structural mechanism. HRTEM images of the formed Fe atomic chain with three and one atom in cross-section are shown in Fig. 4.5g, h, and the schematic of the final structure is demonstrated in Fig. 4.5i, with the inter-atomic distances marked. It was found that the inter-atomic distances were ~2.2 Å, larger than those of bulk crystals of 2.0 Å.

To summarize, the rarely studied BCC structured Fe MACs have the following structural features:

1. The atomic distances are larger than those of bulk crystals. Distortions and slips were observed during the formation processes. This feature was also observed for the FCC metals [18, 19, 23, 24].
2. The surface of the Fe atomic chain is composed of (110) and (001) planes, whereas the surfaces of FCC MACs are mainly composed of (111) planes [14, 17, 25, 26].

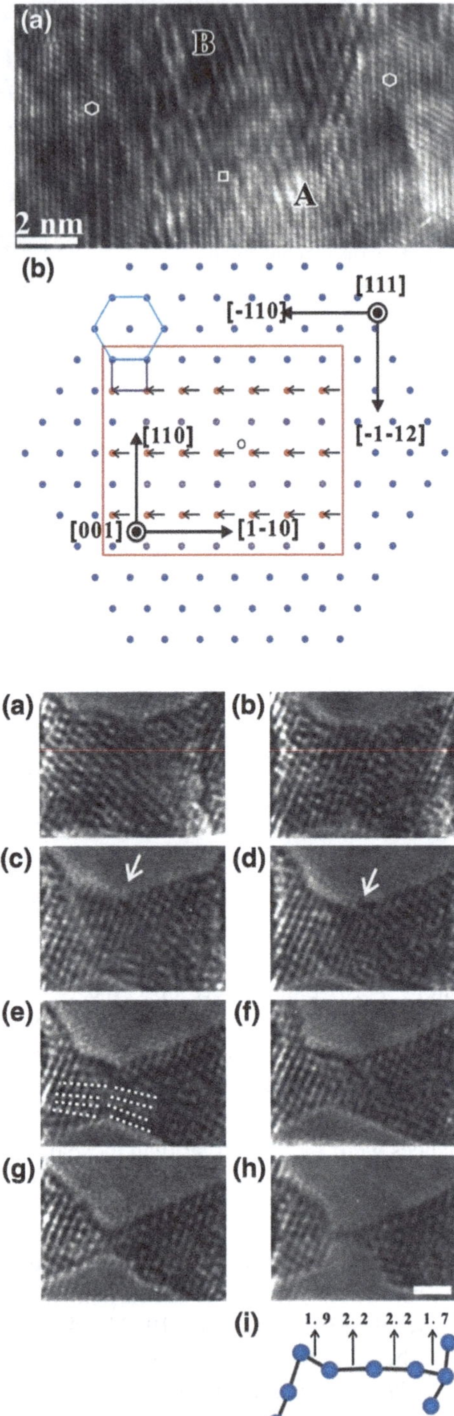

Fig. 4.4 HRTEM image (**a**) and schematic (**b**) of the reorientation from [111] zone axis to [001] zone axis of the single-crystalline Fe during the thinning process

Fig. 4.5 **a–h** HRTEM images showing the formation process of a Fe atomic chain. Atomic steps can be observed at the surface (marked by *arrows* in **c** and **d**). Distortion along (110) is marked with *dotted lines* in **e**. The *scale bar* is 2 nm. **i** Schematic drawing of the Fe atomic chain shown in **h**, the projected inter-atom distances are marked in Å [22]

3. When the thickness of the metals is thin enough, reconstructions from [111] to [001] zone axis was observed for Fe atomic chain, whereas opposite transition was reported for the FCC metals [14, 17, 25].

The distinct deformation behaviors of Fe atomic chains could be attributed to the different atomic structure and surface properties of BCC metals compared with FCC metals. For FCC metals, (111) is the most stable plane, while for BCC metals (110) is the highest densely packed plane. When the thickness of the metals decreases to a threshold value, driven by surface tension, crystals will tend to reorient to the most stable structure, because of the larger ratio of surface atoms compared with the bulk crystals [14, 17, 25, 27, 28].

4.2.2 Formation Mechanism of Fe Atomic Chains

In addition to the above HRTEM observations, the formation process of the Fe atomic chains was investigated by first principles calculations. The structural evolutions and energy changes were simulated and calculated to complement the in situ results. As shown in Fig. 4.6, a Fe nanowire oriented along [110] with 5 atomic layers was built as the starting model. Uniaxial elongation was carried out with a step of 0.4 Å. At first, symmetric deformation was observed by pulling out the center atoms at the second and forth layers towards the midpoint of the nanowire (a-II). Correspondingly, the total energy was increased (b). Under further elongation, an octahedral was formed at the center of the nanowire, along with a local peak of the total energy. As the elongation continued, the deformation became asymmetric (a-IV) and the nanowire was distorted along the (110) plane, along with a decrease of the total energy. Therefore, it was confirmed that distortion was the plastic deformation mechanism, to release the stress during the elongation. Finally, the total energy increased along with further elongation and reached the maximum value when the atomic chain was formed (V–VII). The interatomic

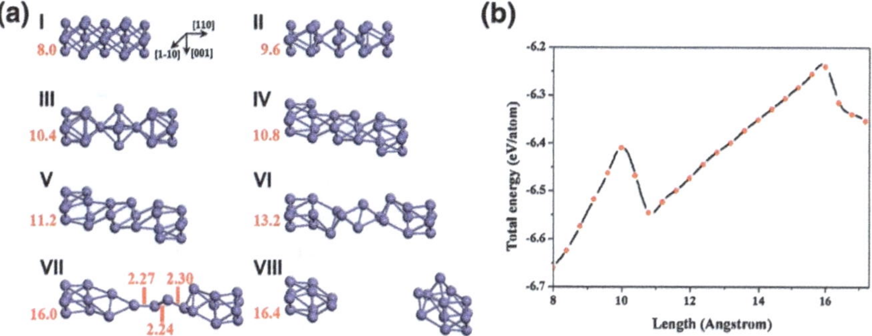

Fig. 4.6 Structural simulations (**a**) and total energy calculations (**b**) during the formation process of a Fe atomic chain by using density functional theory (DFT) approach [22]

distances of the atomic chain were measured to be 2.24–2.30 Å, which is in good agreement with the HRTEM results.

4.3 Electronic Structures of CNT-Clamped Fe Atomic Chains

In addition to the formation mechanisms studied by using HRTEM and theoretical calculations, the electronic structures of CNT-clamped Fe atomic chains were investigated by first principles calculations as demonstrated in Fig. 4.7. A locally stable configuration during the structural evolutions (Fig. 4.6a-III) was selected to connect to two (5, 5) single-walled CNTs. The fully optimized structure is shown in Fig. 4.7a-I. The CNT was found to shrink to match the size of the Fe atomic chain, indicating the strong interactions between the two components. The average length of the Fe–C bonds was measured to be 2.06 Å, quite close to the value of bulk Fe_3C crystal [29]. Spin-resolved density of states (DOS) was calculated as presented in Fig. 4.7Aii–IV, along with the local DOS of Fe and CNT, respectively. From the total DOS, the CNT-clamped Fe atomic chain was metallic and spin-polarized, with spin-up state to be the major state. From the local DOS of CNT, the CNT is metallic with symmetric distributions for both spin states. In contrast, Fe atomic chain is half metallic with the electron state only in one spin state and behaviors like a semiconductor for the spin-down state with a band-gap of ~0.5 eV. Therefore, after the combination with CNT, Fe atomic chain could keep the unique physical properties. The charge distribution of the CNT-clamped Fe atomic chain is demonstrated in Fig. 4.7b. For both cross-sections along [100] and [010], we can see covalent bonding with directional distribution of the charges between CNT and Fe, consistent with previous studies [11].

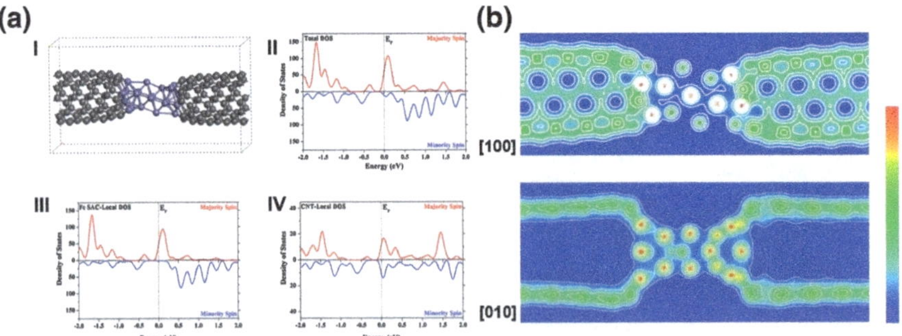

Fig. 4.7 DFT calculations on electronic properties of CNT-clamped Fe atomic chain: **a** *I* Fully relaxed ball-and-stick model of a CNT-clamped Fe atomic chain. *II* Spin resolved total DOS for the CNT-clamped Fe atomic chain. *III, IV* Spin resolved local DOS of the Fe atomic chain and CNT. **b** Charge density of two slices cut along [100] and [010] [22]

4.4 In Situ Measurements of the Electrical Properties of CNT-Clamped Fe Atomic Chains

As mentioned in the introduction of this chapter, one of the unique properties of MACs is the quantized conductance. In this section, the electrical transport properties were measured in situ by using a TEM-STM holder. The experiment procedures included the production of Fe-filled CNTs, dispersion of the sample, and loading of the single CNTs on to Au electrodes, and manipulation of the STM tip to contact the filled CNT. And then fabrication of CNT-clamped Fe atomic chain was carried out similarly to previous methods, except for the final stage where a STM tip was used to apply tensile stress, and also the current–voltage signals were recorded.

4.4.1 Annealing and Its Effect on the Electrical Properties of the Fe-Filled CNTs

To get the true electrical properties, it is necessary to anneal the system by applying a constant bias and current (<10 μA). The annealing process also helps to remove the absorbed contamination to avoid possible influences on the measurements. During the annealing process, electrical properties were monitored to check the contact conditions as shown in Fig. 4.8. Initially, only noise around 2 nA could be measured (Fig. 4.8a), suggesting an insulating layer at the contact. During the annealing (Fig. 4.8b), the contact resistance was lowered and the current increased along with the voltage. A non-linear feature was observed for the second stage, which could be attributed to the Schottky contact. After the annealing procedure (Fig. 4.8c, d), linear feature of the I–V curve was obtained, indicating the Ohmic contact. And the resistance was about 10 kΩ, quite close to the resistance corresponding to quantum conductance (~13 kΩ). The resistance of the circuit could be used for the calculations of the conductance of the CNT-clamped MACs. In addition, the zero point of the voltage was also calibrated to be ~18 mV. Another result of the annealing procedure is the automatic welding of the CNT with the STM tip due to the local Joule heating, which would facilitate the application of uniaxial tension in producing MACs.

4.4.2 Quantized Conductance of CNT-Clamped Fe Atomic Chains

The electrical measurements were conducted during the fabrication of CNT-clamped Fe atomic chains by using a TEM-STM in situ holder, as demonstrated in Fig. 4.9a. After the removal of carbon shells and thinning of the exposed nanowire,

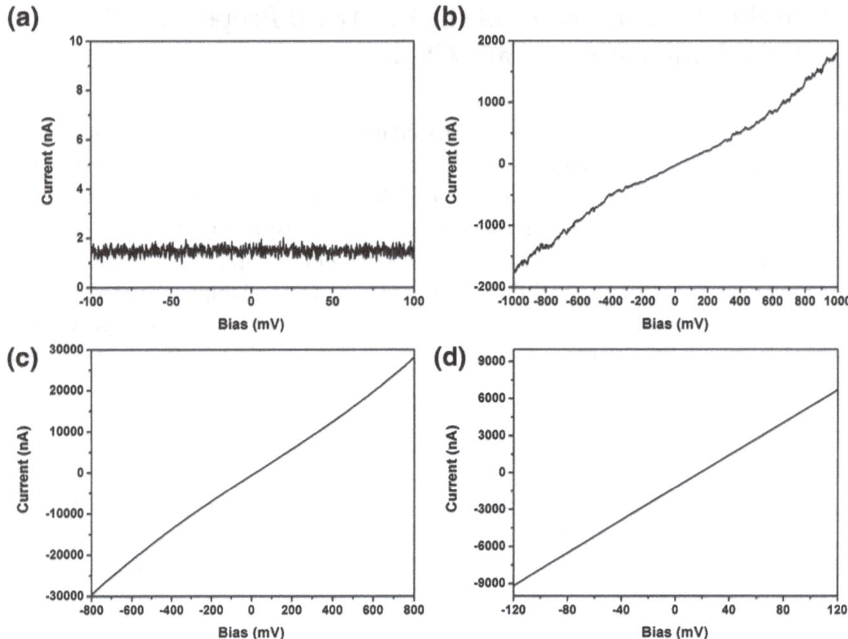

Fig. 4.8 Current–voltage curves of a CNT-clamped Fe nanowire, before (**a**), during (**b**, **c**) and after (**d**) the annealing process, revealing the decreasing of electrical resistance, and change from insulating to Ohmic contact [22]

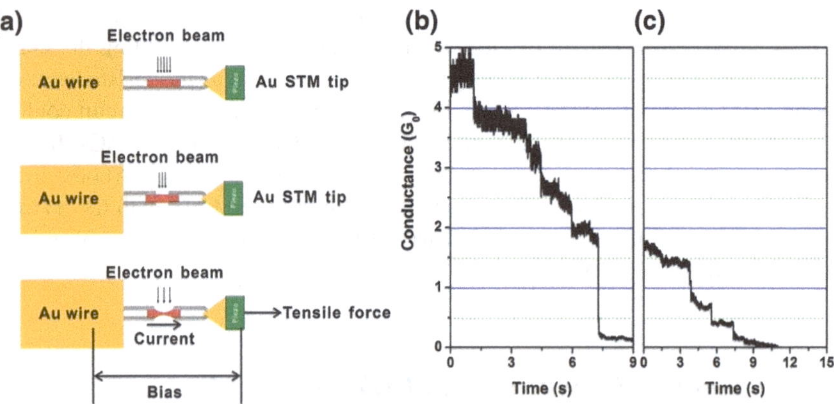

Fig. 4.9 In situ electrical transport measurements of a CNT-clamped Fe atomic chain: **a** Schematic diagram of the experimental process, **b**, **c** Conductance changes as a function of time for two formation and thinning processes of Fe atomic chains at a constant bias of 12 mV, where the contact ruptured comparatively abruptly in (**b**) and gently in (**c**) [22]

the STM tip welded with the CNT was retracted with a very slow speed (~0.1 nm/s), and the conductance was recorded and calculated simultaneously, by using the following formula, $G = [(V - V_0)/I - R_0]^{-1}/G_0$, where G is the conductance, I and V are the measured current and bias, V_0 and R_0 are the corrected system bias and resistance, and G_0 is the quantum conductance $G_0 = 2e^2/h$ ~13 $k\Omega^{-1}$.

The conductance changes along with the formation of a Fe atomic chain is shown in Fig. 4.9b, by keeping the applied bias at 12 mV. It is clearly revealed that the conductance was decreased in a stepwise manner. Importantly, the heights of the steps were on the multiples of 0.5 G_0, such as 2, 2.5, 3.5 and 4.5. When the atomic chain was fractured, it could be reconnected by approaching the ends. The conductance was recovered to be 2 G_0. After that, a slower rate elongation was applied to draw the new atomic chain; steps corresponding to 0.5 and 1.5 G_0 were identified (Fig. 4.9c).

Lots of publications have reported the electrical properties of MACs [1–4, 21, 30–32]. Landau equation was applied to explain the electrons transfer through the small structures with the length shorter than the electrons scattering free path. $G = G_0\Sigma T_i$, where $G_0 = 2e^2/h$, and T_i is the electron transmittance of the *i*th channel [20]. The electrical properties of the atomic chains are mainly determined by the electrons near the Fermi level. For non-magnetic metals such as Au and alkali metals, the conductance of the MACs are multiples of G_0 [2], whereas for the magnetic metals, the conductance locates at multiples of 0.5 G_0 due to the spin-polarization [32]. Therefore, our in situ measurements are consistent with the theories. It is worth mentioning that besides the plateaus at multiples of 0.5 G_0, there are some conductance steps located at other positions such as 0.8 G_0. The reason for the inconsistence could be ascribed to the free electrons approximation of Landau theory, which is more suitable for the simple metals such as alkali metals. For metals with complicated electronic structures such as Fe, it is reasonable to have the electron transmittance equal to fractions of G_0.

4.4.3 Current–Voltage Characteristics of CNT-Clamped Fe Atomic Chains

The structure and electrical conductance of the MACs are quite stable in the range of several seconds. Therefore, it is possible to suspend the structural deformation and study the current–voltage (I–V) characteristics. TEM images and corresponding I–V curve are shown in Fig. 4.10. After the electron irradiation for carbon removal, the exposed Fe nanowire was thinned. Because of the high precision of electron beam, the irradiated area was limited to a range of ~60 nm and leaving other areas unaffected (a). After the electron thinning and elongation by STM, a dual-conical shape was formed with the smallest part around 0.75 nm in diameter, corresponding to 3–4 Fe atoms in cross-section. The I–V curve (c) shows a linear feature, with the corresponding conductance to be ~1.5 G_0.

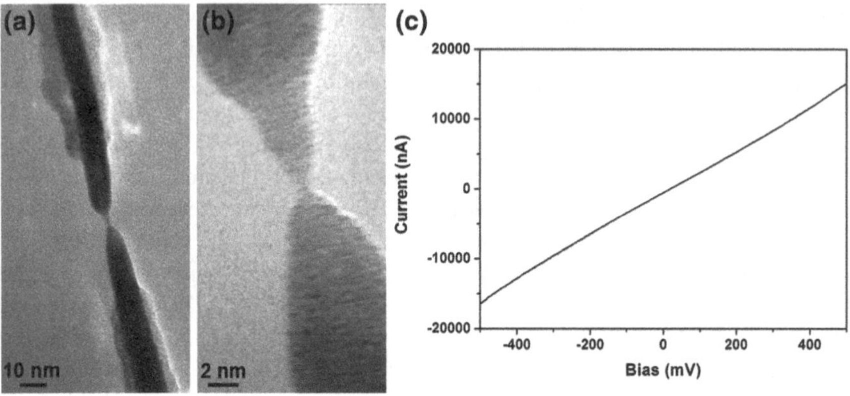

Fig. 4.10 Electrical conductance measurement of a CNT-clamped Fe MAC: **a**, **b** Low and high magnification TEM images, **c** Current–voltage curve [22]

Mehrez et al. reported that the nonlinear I–V characteristics are caused by the adsorption of gases [21, 32, 33]. In our work, the linear I–V feature indicates that the contacts between at the boundaries of CNT and MAC are Ohmic and there is little gas absorbed on the surface. The clean surface could be attributed to the high vacuum inside the TEM, avoiding the use of organic materials like glues, and the effective annealing processes.

4.5 CNT-Clamped MACs of Other Metals

In previous sections, mainly the CNT-clamped Fe atomic chains were studied. Fe is a BCC structured and carbide formation metal, and also a catalyst for CNTs growth. The possibilities of using CNTs for connection of other MACs were investigated in this section. These metals include Fe–Ni alloy, non-catalyst and non-carbon formation metal Pt with FCC structure, and Co with hexagonal close-packed (HCP) structure.

4.5.1 Fabrication of CNT-Clamped MACs of Fe Alloy and Pt

A key step to obtain the CNT-clamped MACs is the continuous filling of metals in CNTs. Similar to the Fe filling, Fe alloys could be filled into CNTs by using floating catalyst CVD method with the mixture of ferrocene, nickelocene, and cobaltocene as the catalyst precursor. And as shown in Fig. 4.11, Fe–Ni atomic chains could be fabricated with the same technique. It is interesting to mention

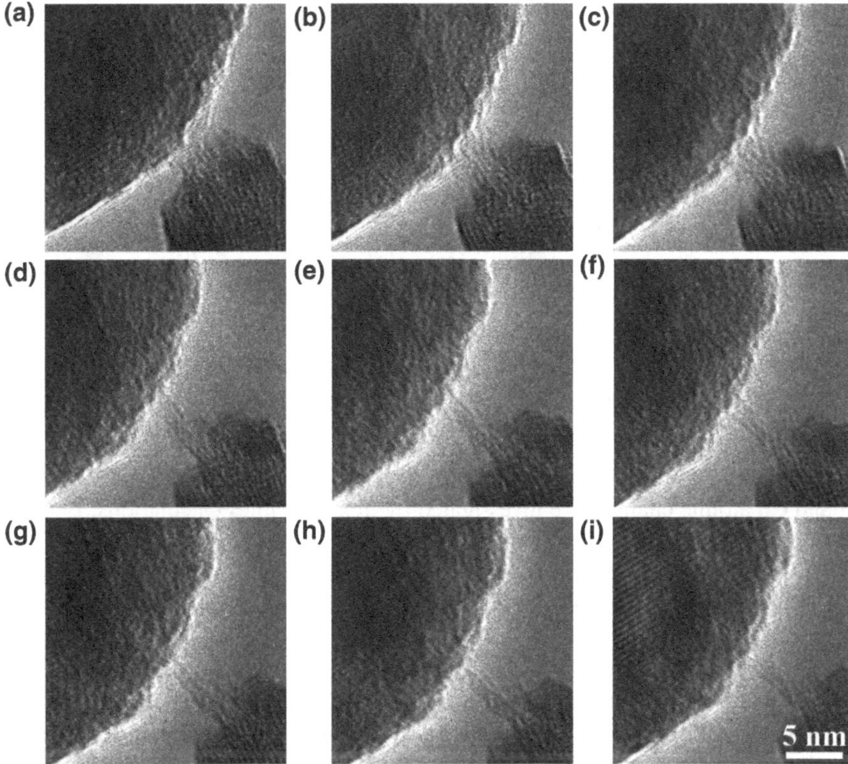

Fig. 4.11 TEM observations of the formation process of Fe–Ni alloy MAC [22]

that before the fracture of the formed atomic chain, the length increased from 1.5 to 5.7 nm, and the diameter shrank from 4.25 to 0.53 nm, indicating the larger plasticity of the alloy atomic chains than that of pure Fe atomic chains.

The filling of CNTs by using floating catalyst CVD method is limited to the catalyst metals for CNTs growth. To prepare CNTs with other metals, a template method was adopted [34]. Firstly, CNTs were deposited into the channels of AAO templates by thermal decomposition of hydrocarbon precursors. After that, metals were filled into the CNTs by electrochemical deposition method. Similar to the fabrication processes of CNT-clamped MACs of catalyst metals, electron beam irradiation was applied to peel off the carbon shells and thin the exposed metal nanowires. The formation process of a Pt atomic chain is demonstrated in Fig. 4.12. Initially, the Pt nanowire was highly crystalline with well resolved lattices. After the removal of carbon shells, electron beam was focused on the lower part of the nanowire (a) and then the upper part (b). Gradually, the exposed Pt nanowire was thinned layer by layer until the formation of atomic chains with two or only one atom in the cross-section (d, e).

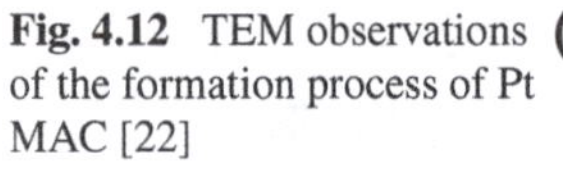

Fig. 4.12 TEM observations of the formation process of Pt MAC [22]

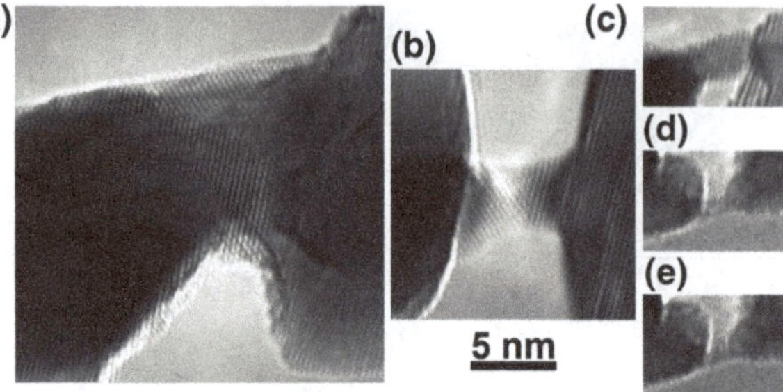

4.5.2 *First Principles Calculations of CNT-Clamped Pt, Co Atomic Chains*

The structure and electronic properties of the hybrid structures between CNTs and Pt and Co atomic chains were investigated by first principles calculations. Because Pt is a typical FCC metal and Co is HCP structured, the axes were set to be along [111] and [002] directions, respectively, as demonstrated in Figs. 4.13 and 4.14. For both metals, uniaxial tension was applied and it was revealed that before the fracture of the nanowires, atomic chains with only one atom in width were formed, consistent with previous reports [1, 5, 35, 36]. CNT-clamped atomic chains of Pt and Co were built on the basis of the results shown in Fig. 4.13g and Fig. 4.14g. After the structural optimization, the atomic chains were shrunk, and covalent bonding was formed at the boundaries as demonstrated by the charge distributions.

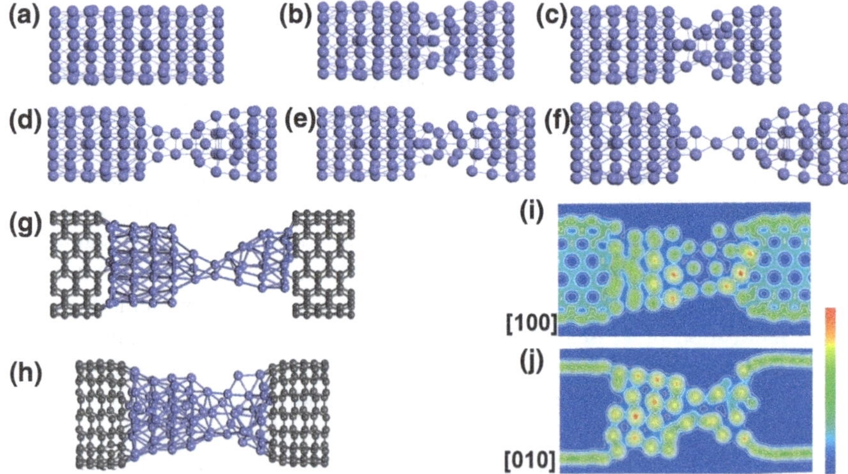

Fig. 4.13 DFT study of a CNT-clamped Pt atomic chain: **a–f** Simulation on the structure evolution of the Pt atomic chain, **g, h** Ball-and-stick model of a CNT-clamped Pt atomic chain before and after full relaxation, **i, j** Charge density after structural relaxation of two slices cut along [100] and [010] [22]

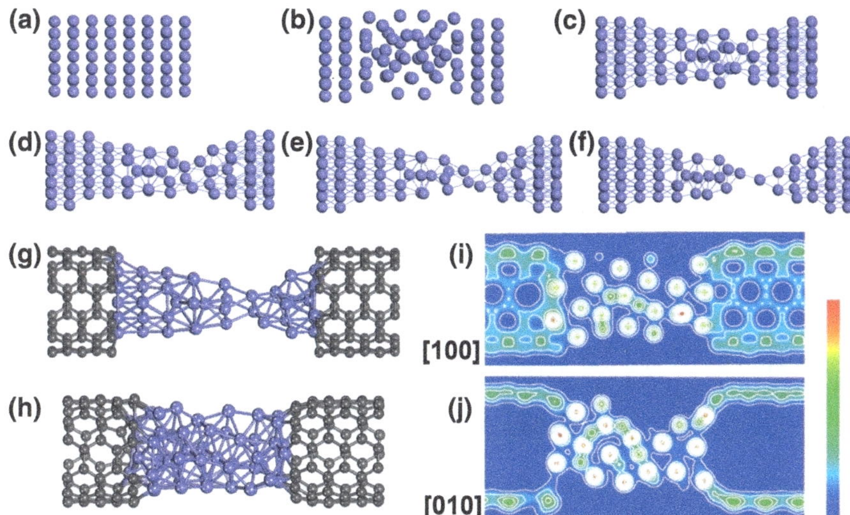

Fig. 4.14 DFT study of a CNT-clamped Co atomic chain. **a–f** Simulation on the formation process of the Co atomic chain, **g**, **h** Ball-and-stick model of a CNT-clamped Co atomic chain before and after full relaxation, **i**, **j** Charge density after structural relaxation of two slices cut along [100] and [010] [22]

Above results demonstrate that MACs of various metals including the catalysts for CNTs growth or non-catalysts, carbide formation or non-formation metals, and three different crystalline structures, could be connected effectively by CNTs. Therefore, our method is universal for the connection of MACs.

4.6 Summary

CNT-clamped MACs were designed and fabricated by in situ TEM method. Combined with first principles calculations, the formation mechanism and electrical properties were investigated, and the following conclusions are drawn:

1. CNT-clamped MACs is proposed for the effective connection and assembly of the extremely small MACs. This new concept may facilitate applications of MACs in nano electronic and spintronic devices.
2. The formation process is investigated by in situ TEM and first principles calculations. It is revealed that slip and twist along (110) are the important mechanisms. In addition, it is found that the surface of Fe atomic chains are composed of (110) planes which is different from the (111) planes of FCC metals, due to the different atomic arrangement and surface properties.
3. The electronic structures of CNT-clamped MACs are studied by first principles calculations. The two components are bonded by strong covalent bonding.

Pristine properties such as the half-metallicity of Fe atomic chains and metallicity of CNTs remain unchanged after the combination. In addition, the electrical properties are measured in situ and found to be quantized.

4. The fabrication technique is applied to other metals and alloys, and CNTs-clamped atomic chains of Fe–Ni alloy and Pt are produced successfully. First principles calculations demonstrate that these atomic chains of BCC structured Fe, FCC structured Pt and HCP structured Co could be effectively connected with CNTs.

References

1. Agraït N, Yeyati AL, van Ruitenbeek JM (2003) Quantum properties of atomic-sized conductors. Phys Rep 377:81–279
2. Krans JM, Vanruitenbeek JM, Fisun VV, Yanson IK, Dejongh LJ (1995) The signature of conductance quantization in metallic point contacts. Nature 375:767–769
3. Ohnishi H, Kondo Y, Takayanagi K (1998) Quantized conductance through individual rows of suspended gold atoms. Nature 395:780–783
4. Yanson AI, Bollinger GR, van den Brom HE, Agraït N, van Ruitenbeek JM (1998) Formation and manipulation of a metallic wire of single gold atoms. Nature 395:783–785
5. Sokolov A, Zhang CJ, Tsymbal EY, Redepenning J, Doudin B (2007) Quantized magnetoresistance in atomic-size contacts. Nat Nanotechnol 2:171–175
6. Smith DPE (1995) Quantum point-contact switches. Science 269:371–373
7. Baughman RH, Zakhidov AA, de Heer WA (2002) Carbon nanotubes—the route toward applications. Science 297:787–792
8. Frank S, Poncharal P, Wang ZL, de Heer WA (1998) Carbon nanotube quantum resistors. Science 280:1744–1746
9. Tans SJ, Verschueren ARM, Dekker C (1998) Room-temperature transistor based on a single carbon nanotube. Nature 393:49–52
10. Bachtold A, Hadley P, Nakanishi T, Dekker C (2001) Logic circuits with carbon nanotube transistors. Science 294:1317–1320
11. Rodríguez-Manzo JA, Banhart F, Terrones M, Terrones H, Grobert N, Ajayan PM, Sumpter BG, Meunier V, Wang M, Bando Y, Golberg D (2009) Heterojunctions between metals and carbon nanotubes as ultimate nanocontacts. Proc Natl Acad Sci USA 106:4591–4595
12. Tsukagoshi K, Alphenaar BW, Ago H (1999) Coherent transport of electron spin in a ferromagnetically contacted carbon nanotube. Nature 401:572–574
13. Agraït N, Rodrigo JG, Vieira S (1993) Conductance steps and quantization in atomic-size contacts. Phys Rev B 47:12345–12348
14. Kondo Y, Takayanagi K (1997) Gold nanobridge stabilized by surface structure. Phys Rev Lett 79:3455–3458
15. Muller CJ, van Ruitenbeek JM, de Jongh LJ (1992) Conductance and supercurrent discontinuities in atomic-scale metallic constrictions of variable width. Phys Rev Lett 69:140–143
16. Kondo Y, Takayanagi K (2000) Synthesis and characterization of helical multi-shell gold nanowires. Science 289:606–608
17. Kondo Y, Ru Q, Takayanagi K (1999) Thickness induced structural phase transition of gold nanofilm. Phys Rev Lett 82:751–754
18. Rodrigues V, Ugarte D (2001) Real-time imaging of atomistic process in one-atom-thick metal junctions. Phys Rev B 63:073405
19. Rubio G, Agraït N, Vieira S (1996) Atomic-sized metallic contacts: mechanical properties and electronic transport. Phys Rev Lett 76:2302

20. Landauer R (1970) Electrical resistance of disordered one-dimensional lattices. Philos Mag 21:863
21. Chopra HD, Sullivan MR, Armstrong JN, Hua SZ (2005) The quantum spin-valve in cobalt atomic point contacts. Nat Mater 4:832–837
22. Tang D-M, Yin L-C, Li F, Liu C, Yu W-J, Hou P-X, Wu B, Lee Y-H, Ma X-L, Cheng H-M (2010) Carbon nanotube-clamped metal atomic chain. Proc Natl Acad Sci USA 107:9055–9059
23. Rodrigues V, Fuhrer T, Ugarte D (2000) Signature of atomic structure in the quantum conductance of gold nanowires. Phys Rev Lett 85:4124–4127
24. Rego LGC, Rocha AR, Rodrigues V, Ugarte D (2003) Role of structural evolution in the quantum conductance behavior of gold nanowires during stretching. Phys Rev B 67:045412
25. Diao JK, Gall K, Dunn ML (2003) Surface-stress-induced phase transformation in metal nanowires. Nat Mater 2:656–660
26. Bettini J, Sato F, Coura PZ, Dantas SO, Galvão DS, Ugarte D (2006) Experimental realization of suspended atomic chains composed of different atomic species. Nat Nanotechnol 1:182–185
27. Zhang JM, Ma F, Xu KW (2004) Calculation of the surface energy of FCC metals with modified embedded-atom method. Appl Surf Sci 229:34–42
28. Zhang J-M, Wang D-D, Xu K-W (2006) Calculation of the surface energy of bcc transition metals by using the second nearest-neighbor modified embedded atom method. Appl Surf Sci 252:8217–8222
29. Wood IG, Vocadlo L, Knight KS, Dobson DP, Marshall WG, Price GD, Brodholt J (2004) Thermal expansion and crystal structure of cementite, Fe_3C, between 4 and 600 K determined by time-of-flight neutron powder diffraction. J Appl Crystallogr 37:82–90
30. Brandbyge M, Schiøtz J, Sørensen MR, Stoltze P, Jacobsen KW, Nørskov JK, Olesen L, Laegsgaard E, Stensgaard I, Besenbacher F (1994) Quantized conductance in an atom-sized point contact. Phys Rev Lett 72:2251
31. de Picciotto R, Stormer HL, Pfeiffer LN, Baldwin KW, West KW (2001) Four-terminal resistance of a ballistic quantum wire. Nature 411:51–54
32. Komori F, Nakatsuji K (1999) Quantized conductance through atomic-sized iron contacts at 4.2 K. J Phys Soc Jpn 68:3786–3789
33. Mehrez H, Wlasenko A, Larade B, Taylor J, Grütter P, Guo H (2002) I–V characteristics and differential conductance fluctuations of Au nanowires. Phys Rev B 65:195419
34. Kyotani T, Pradhan BK, Tomita A (1999) Synthesis of carbon nanotube composites in nanochannels of an anodic aluminum oxide film. Bull Chem Soc Jpn 72:1957–1970
35. Rodrigues V, Bettini J, Silva PC, Ugarte D (2003) Evidence for spontaneous spin-polarized transport in magnetic nanowires. Phys Rev Lett 91:096801
36. Smit RHM, Untiedt C, Yanson AI, van Ruitenbeek JM (2001) Common origin for surface reconstruction and the formation of chains of metal atoms. Phys Rev Lett 87:266102

Chapter 5
Conclusions and Perspective

5.1 Conclusions

The growth mechanism, controllable synthesis, and practical applications are the most important topics for current researches of CNTs. The fundamental understanding of the growth mechanism is the key to achieving controllable synthesis. And selective synthesis is the foundation for practical applications in nanoelectronic devices. At the same time, the potential applications also provide targets for the control of synthesis. In this thesis, in situ TEM techniques are employed to investigate the growth mechanism and CNT-based device fabrications. The main conclusions are as follows:

1. The nucleation processes of CNTs from metal-free SiO_x catalysts were observed for the first time. It was revealed that the active species of the catalyst is silicon oxide in solid and amorphous states, rather than Si or SiC. The growth of CNTs from SiO_x catalyst follows a vapor–solid surface–solid mechanism.
2. The states of Fe and SiO_x catalysts during CNTs nucleation were investigated and compared by using in situ TEM. The general requirements for the catalyst were proposed, including the suitable size of catalyst particles, suitable interactions between the catalyst and the carbon, and necessary driving force. And the key points in conventional vapor–liquid–solid growth mechanisms such as the catalytic ability to decompose the hydrocarbon gases, liquid carbide phase, and dissolution-over saturation–precipitation route were found to be not indispensable.
3. The structural evolutions of carbon microcoils under high-density current were investigated by using in situ TEM. It was revealed that in the absence of catalyst, double-walled CNTs could be obtained by the reconstruction of carbon structures, confirming the possibility of catalyst-free growth of CNTs.
4. Based on the understanding of growth mechanism, a heteroepitaxial growth approach was developed by using boron nitride platelet nanofibers as the growth seeds for growing CNTs. As a result, single-walled CNTs with the diameters distributed on multiples of (002) interplanar distances were produced, providing a new approach for the controllable synthesis.

D.-M. Tang, *In Situ Transmission Electron Microscopy Studies of Carbon Nanotube Nucleation Mechanism and Carbon Nanotube-Clamped Metal Atomic Chains*, Recognizing Outstanding Ph.D. Research, DOI: 10.1007/978-3-642-37259-9_5,

5. A CNT-clamped metal atomic chain (MAC) hybrid structure was designed and fabricated by using in situ TEM method for the connection of extremely small MACs. The formation mechanism of CNT-clamped Fe atomic chain (AC) was investigated by means of in situ HRTEM and first principles calculations. It was demonstrated that distortion and slip along (110) plane was the deformation mechanism. The surfaces of Fe AC were composed of (110) and (001) planes, distinctive from the surfaces of FCC metals, which were composed of (111) planes.
6. First principles calculations indicated that CNTs can form strong covalent bonding with Fe ACs. And Fe ACs could well retain the unique physical properties after the combination with CNTs. In addition, in situ electrical measurements demonstrated that the conductance of the CNT-clamped Fe AC was quantized.

5.2 Perspective for Future Works

1. In addition to the in situ observations of the catalyst states during the CNTs nucleation in this thesis, it is necessary to further investigate the growth mechanism by using environmental TEM to correlate the CNT diameters and chirality with the catalyst composition, atmosphere, and other growth parameters.
2. An effective approach for the precise control of CNTs structures was developed in this thesis by using high-temperature stable boron nitride nanofibers as growth seeds. However, the growth rate was still not high enough. Therefore, activation of the boron nitride should be explored to enhance the productivity. New growth seeds should be designed following the philosophy to control the structure of single-walled CNTs.
3. Our work has demonstrated the advantage of using CNTs for nanoscale connections of extreme small structures, for the nanometer size, low scattering of electron and spin transport. However, the present work using in situ TEM is not practical for real applications. Therefore, it is expected that other techniques such as lithography will be used for the practical fabrications of nanodevices such as CNT-clamped MACs.

Zeitfracht Medien GmbH
Ferdinand-Jühlke-Straße 7
99095 Erfurt, Deutschland
produktsicherheit@kolibri360.de